KU-510-577

Arid Environments

Lucy Cole

Advanced TopicMaster

Series editor
Michael Raw

Philip Allan Updates, an imprint of Hodder Education, an Hachette UK company, Market Place, Deddington, Oxfordshire OX15 0SE

Orders

Bookpoint Ltd, 130 Milton Park, Abingdon, Oxfordshire OX14 4SB
tel: 01235 827720
fax: 01235 400454
e-mail: uk.orders@bookpoint.co.uk
Lines are open 9.00 a.m.–5.00 p.m., Monday to Saturday, with a 24-hour message answering service. You can also order through the Philip Allan Updates website: www.philipallan.co.uk

© Philip Allan Updates 2010

ISBN 978-1-44410-834-7

First printed 2010
Impression number 5 4 3 2 1
Year 2014 2013 2012 2011 2010

All rights reserved; no part of this publication may be reproduced, stored in a retrieval system, or transmitted, in any form or by any means, electronic, mechanical, photocopying, recording or otherwise without either the prior written permission of Philip Allan Updates or a licence permitting restricted copying in the United Kingdom issued by the Copyright Licensing Agency Ltd, Saffron House, 6–10 Kirby Street, London EC1N 8TS.

Printed in Spain

Hachette UK's policy is to use papers that are natural, renewable and recyclable products and made from wood grown in sustainable forests. The logging and manufacturing processes are expected to conform to the environmental regulations of the country of origin

P01681

Contents

Overview

Process

Form

Contents

Introduction

This book provides a detailed and up-to-date review of arid environments for AS/A2 geography students. It is aimed primarily at sixth-form students, but the content of the book may also be useful for first-year undergraduates.

Around one-third of the Earth's land surface is classified as either arid or semi-arid. The media provide us with many images. Television, films, newspapers and the internet often portray these lands as hot, dry, barren areas of sand dunes, either uninhabited or where life is a constant struggle for survival. The reality, however, is degrees of aridity, encompassing the utmost arid regions at one extreme and semi-arid areas at the other, and with large variations in climate, water resources, landscapes and ecosystems.

Arid environments are not just of interest to academics. These lands are home to almost one-fifth of the world's population. Today, deserts are areas of economic development supporting industries such as agriculture, tourism, mineral extraction and film-making. Yet arid lands are arguably the harshest of natural environments, and have a water deficit for most of the year. The consequent over-exploitation of water resources leads to environmental and political problems.

Desertification is one of the most serious environmental issues of today. Increasing population pressure is leading to land degradation and the spread of arid lands. Global warming is likely to exacerbate this problem and increase the amount of land at risk. Arid lands are fragile. If these environments are to be managed and used in a sustainable manner, an understanding of the processes operating here is crucial.

Students can use this book in several ways. Most obviously, the material provides a general understanding of processes, forms and human interactions in arid environments. The text should be read to consolidate students' learning as each topic is covered in class. Specific areas of the text can be used to complete essays and other assignments. The diagrams, tables and photographs should be studied carefully and used to illustrate and extend the concepts and facts covered by the text. General discussion and explanation is backed up by case studies, which can be used to support assignments and written work in the examinations. Activities interspersed throughout the text aim to deepen your knowledge and understanding, to develop skills, including the statistical analysis of data, and to encourage you to investigate topics further.

Lucy Cole

1 An overview of arid and semi-arid environments

Why study arid and semi-arid environments?

Arid and semi-arid environments cover around a third of the Earth's land surface. The nature of these environments has a strong impact on:
- plant and animal life
- geomorphological processes and landscape development
- the global distribution of people and levels of development in these regions

Myths abound about the arid zone — far from the simplistic media view of hot, dry, unvegetated and uninhabited sand dunes, the arid environment is characterised by considerable diversity. An estimated 14% of the world's population live in arid and semi-arid environments; while life in some arid areas may be a constant struggle to tame the land, it is not always a case of economic hardship, drought and famine.

Arid and semi-arid environments present a range of opportunities and challenges for human activity. How people live within these areas depends on their perception of the land and their understanding of the fragility of the environment. It is well known that human activity can lead to increased problems in arid areas, and that a rising population will generate increased demands and put more pressure on arid and semi-arid lands. Therefore it is of the utmost importance that we should understand the human impact on arid and semi-arid environments in order to best understand how people should manage these areas in the future.

Activity 1

Use the internet to research:

(a) the opportunities for human activity in arid and semi-arid environments

(b) the challenges that arid and semi-arid environments are likely to face in the future

Activity 1 (continued)

The BBC news website pages listed below are a good starting point for this exercise.

news.bbc.co.uk/1/hi/sci/tech/7456973.stm

news.bbc.co.uk/1/hi/sci/tech/5041988.stm

news.bbc.co.uk/1/hi/health/443247.stm

Defining arid and semi-arid environments

Arid and semi-arid environments are those that lack moisture and do not have sufficient rainfall to support most trees or woody plants. The climates of these areas are extremely dry, with low levels of precipitation and high rates of **evapotranspiration** (the sum of evaporation and plant transpiration from the Earth's surface to the atmosphere). The scarcity of moisture is the main factor limiting biological processes.

Aridity defined by annual rainfall

The simplest definition of aridity is in terms of total annual precipitation. Table 1.1 shows how precipitation can be used to group areas according to whether they are hyper-arid, arid or semi-arid. For comparison, in the UK, parts of the Western Highlands of Scotland receive on average over 5000 mm of rain a year, and the driest parts of East Anglia receive about 500 mm.

There are, however, problems with the classification in Table 1.1. It does not take account of extreme events, or consider the type, frequency, intensity and seasonal distribution of precipitation. Precipitation in arid and semi-arid environments occurs infrequently, often in heavy downpours, and in the driest areas it is erratic and unreliable. Therefore the availability of moisture to plants from rainfall will vary considerably over the course of the year or even over a period of several years. Another problem is that the classification in Table 1.1 does not take account of evapotranspiration — and since it does not balance the water inputs with water outputs, it tells us nothing about the amount of water in the system.

| Table 1.1 | Classification of arid areas according to annual precipitation |

Classification	Annual precipitation
Hyper-arid	< 100 mm
Arid	100–250 mm
Semi-arid	250–500 mm

Index of aridity

The relationship between rainfall and potential evapotranspiration is crucial in providing us with a more accurate measure of how much moisture is available in arid environments.

Potential evapotranspiration is defined as the amount of evaporation that would occur if water were freely available for evaporation all year round. It takes into account atmospheric humidity, solar radiation and the wind. In hot arid and semi-arid environments annual potential evapotranspiration always exceeds annual precipitation. Therefore the volume of water that could theoretically be lost through evaporation and transpiration is invariably greater than the volume of water that is actually available.

The United Nations Environment Programme (UNEP) uses the following definition of aridity:

$$AI = \frac{P}{PET}$$

where

AI = aridity index

P = average annual precipitation

PET = annual potential evapotranspiration

NB: P and PET must be expressed in the same units.

Table 1.2 shows the three categories of arid area defined by the aridity index, and the proportion of the Earth's surface represented in each case. According to the UNEP classification, 37.3% of the Earth can be considered to be arid to some extent — over one-third of the world's land area.

| Table 1.2 | Classification of arid areas according to the aridity index |

Classification	Aridity index	% of global land area
Hyper-arid	< 0.05	7.5
Arid	0.05–0.20	12.1
Semi-arid	0.20–0.50	17.7

Source: UNEP, 1997

Activity 2

(a) Copy Table 1.3 and complete it by calculating the aridity index for each place (the first two have been done for you).

Activity 2 (continued)

Table 1.3 Average precipitation, potential evapotranspiration and population (2006) for selected areas in Australia

Place	Annual precipitation (mm)	Potential annual evapotranspiration (mm)	Aridity index	Population size
Perth	820	2000	0.41	1 445 079
Giles	250	2790	0.09	318
Alice Springs	310	2710		26 486
Oodnadatta	70	2700		277
Telfer	290	3200		464
Cairns	3200	1950		135 856
Darwin	2000	2700		120 900
Halls Creek	650	3350		1211
Birdsville	90	2820		326
Mount Isa	410	2910		24 976

(b) Draw a chart to show the relationship between the aridity index and population size.

(c) Use an appropriate statistical technique to test the strength of the relationship between the aridity index and population size. Comment on your findings.

(d) 'Aridity is an important, but not the defining, factor influencing population density in arid and semi-arid environments.' Discuss.

Distinctions between the hyper-arid, arid and semi-arid areas

The **hyper-arid zone** consists of extremely dry areas where the rainfall is infrequent and rarely exceeds 100 mm per year. There may be no rain at all for periods of several years. Apart from a few scattered shrubs, vegetation is absent in the hyper-arid zone. Because of the lack of adequate moisture and vegetation, the main form of farming is nomadic pastoralism.

Figure 1.1 shows a hyper-arid area on the northern edge of the Sahara Desert

Figure 1.1 A hyper-arid area: the Sahara Desert in Morocco

Lucy Cole

in Morocco. This area is characterised by low and erratic precipitation (0.4 mm of rain on average a year) and an aridity index of 0.02. Sand dunes cover approximately 25% of the Sahara; levels of solar radiation are high; and strong winds cause frequent dust- and sandstorms. In this harsh environment, soils are poorly developed and vegetation cover is sparse.

The **arid zone** also receives infrequent and unreliable rainfall, which rarely exceeds 250 mm a year. Vegetation cover remains sparse, comprising plants such as annual grasses, shrubs and small trees. There is some sedentary pastoral farming, and arable farming is possible where aquifers or perennial streams provide sufficient water for irrigation.

The **semi-arid zone** receives higher and more reliable rainfall. Vegetation consists of a variety of grasses, shrubs and trees. Sedentary agriculture is possible, although irrigation may be required in months with low levels of rainfall.

Figure 1.2 shows a semi-arid area in the High Atlas Mountains of Morocco. Average annual precipitation is 363 mm, and the area has an aridity index of 0.295. The rainfall follows a seasonal pattern and is fairly reliable, particularly during the winter months. The natural river system has been modified by a series of dams, artificial lakes and canals, enabling sedentary agriculture along the riverbanks.

| Figure 1.2 | A semi-arid area: the High Atlas Mountains of Morocco |

Lucy Cole

Location of arid environments

Hot arid and semi-arid environments are found in the tropics and subtropics, between latitudes 20° and 35° north and south of the Equator. The extent of the world's arid areas is shown in Figure 1.3.

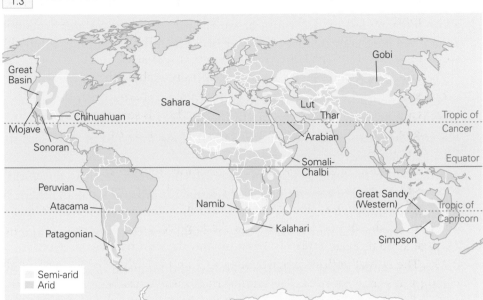

Figure 1.3 Distribution of the hot arid and semi-arid climates, with deserts labelled

The hyper-arid zone extends mainly across the Sahara, Arabian and Gobi deserts in the northern hemisphere, and the Atacama Desert in the southern hemisphere. The arid zone is more extensive, covering much of the area fringing the Sahara, Arabian and Gobi deserts in the northern hemisphere, and much of Australia and southwestern Africa in the southern hemisphere. The semi-arid zone occurs around the poleward margins of the arid zone.

Around 80% of the world's arid and semi-arid areas are contained within just three continents: Africa, Asia and Australasia (Table 1.4). Of the 55 countries in these continents that are to some extent arid, 34 have at least 75% of their land area defined by the UNEP as arid or semi-arid environment.

Table 1.4 Extent of arid areas by continent

Continent	Extent of arid/semi-arid area (million km²)	% of world's arid and semi-arid regions
Africa	17.3	36.7
Asia	15.7	33.3
Australasia	6.4	13.6
North America	4.4	9.3
South America	3.1	6.6
Europe	0.2	0.4

The world's hot deserts

Arid areas can be equated with deserts, since desert areas are considered to be those receiving less than 250 mm of rainfall per year (i.e. arid and hyper-arid areas). We have seen that lack of water is the main factor limiting biological processes.

Major deserts (Figure 1.3) are found in the following continents:

- North America — e.g. the Sonoran and Chihuahuan deserts in the southwest USA and Mexico, and the Mojave in southwest USA
- South America — e.g. the Atacama Desert along the coast of Chile and the Patagonian Desert in Argentina
- Africa — e.g. the Sahara Desert in the north (the largest desert in the world), the Somali–Chalbi Desert in the east and the Namib and Kalahari deserts in southern Africa
- Asia — e.g. the Arabian and Thar deserts astride the Tropic of Cancer and the Lut Desert in eastern Iran
- Australia — the Great Sandy (or Western) and Simpson deserts astride the Tropic of Capricorn

Although all deserts fall into the arid and hyper-arid categories, the aridity index varies in the different desert areas. The most arid deserts are the Saharan and Chilean–Peruvian deserts, followed by the Arabian, east African, Gobi, Australian and southern African deserts. The Thar and North American deserts are generally less arid. The Atacama Desert in Chile, considered the driest place on Earth, gets almost no rainfall. Iquique in the Atacama Desert receives on average just 0.4 mm a year, in the form of rare showers drifting off the Andes Mountains. This can be compared to the Great Basin Desert in North America, where the average annual rainfall varies from 100 mm to 400 mm across the region.

Causes of aridity

There are five major factors causing aridity in hot arid and semi-arid environments: atmospheric pressure, relief, cold ocean currents, continentality and offshore winds. While these factors generally interact, one may be more significant than the others depending on the individual location.

Atmospheric pressure and wind systems

Most of the world's hot arid and semi-arid environments occupy a zone between latitudes 20° and 35° north and south of the Equator. These are zones of persistent dry descending air and stable high-pressure systems. Aridity in these latitudes can be explained by the global atmospheric circulation pattern.

Air is warmed at the Equator and cools at the poles. This creates a pressure gradient that causes the movement of air from high-pressure to low-pressure areas. The global atmospheric circulation pattern is broken up into different cells. It is the low latitude circulation occurring within the two large convective Hadley cells that is important to our understanding of the arid zone (Figure 1.4 and see below).

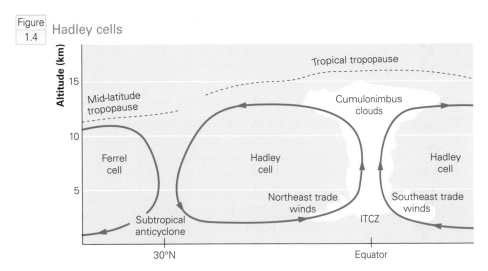

Figure 1.4 Hadley cells

The Hadley and Ferrel cells

- Intense insolation near the Equator leads to surface heating. The air that is heated around the Equator rises through the atmosphere into the troposphere.
- Convection leads to low pressure at the surface and also generates huge cumulonimbus clouds, explaining the ample rainfall found here.
- The warmed air that rises through the troposphere diverges polewards at the tropopause and moves towards higher latitudes.
- As the air moves towards higher latitudes, it cools and sinks back towards the Earth's surface at latitudes 20° to 30°, creating an area of high pressure at the surface. Sinking air is warmed by compression, which prevents cloud formation and rain.
- The circulation is completed by a surface movement of air back towards the Equator (i.e. the trade winds). Surface air converges near to the Equator at the inter-tropical convergence zone (ITCZ).

In temperate latitudes there are comparable circulatory movements of air. Air rises in these latitudes, moves towards the tropics and then descends around latitudes 20° to 30°. These circulatory movements form the Ferrel cells.

As air from both the Hadley and Ferrel cells converges in the sub-tropics, it is forced down towards the Earth's surface and creates an area of high pressure. The sinking air warmed by compression leads to stable conditions. The warming of the air gives it an increasing capacity to hold moisture, making cloud formation unlikely. Subsidence also prevents air from rising from the ground surface and so convection does not take place.

This permanent sub-tropical belt of high pressure is characterised by anticyclonic conditions of clear skies, low amounts of rainfall, high rates of insolation and high rates of evaporation for around 90% of the year. The absence of cloud cover allows a lot of heat to build up during the day, but also for a lot of heat to be lost through radiation at night. The low humidity also promotes heat loss at night. This explains the high diurnal temperature ranges that can be found here.

Extensive continental deserts in the tropics and subtropics are called **climate zone deserts**. Subtropical deserts such as the Sahara, Arabian, Kalahari and Australian deserts are characterised by anticyclonic weather conditions with clear skies, high ground and air temperatures and marked nocturnal cooling. There are seasonal contrasts in conditions but winter temperatures rarely fall as low as freezing and the climate is typically one of hot summers and mild winters.

The impact of the global atmospheric circulation pattern is the main cause of aridity, with other factors having only a local significance.

Relief and the rain shadow effect

High mountain ranges may lead to the formation of desert areas by intercepting precipitation from moist, prevailing winds. In these circumstances, windward slopes of the mountain ranges receive more rainfall than leeward slopes, which will lie in a rain shadow. In extreme cases, the leeward slopes may experience arid or semi-arid climatic conditions.

Figure 1.5 explains the formation of rain shadow deserts. Moist air brought inland by prevailing winds is forced to rise over the mountains. As it does so, it cools and expands. Condensation occurs on the windward side of the mountains, since the colder air cannot hold as much water vapour as the warm air. The result is cloud formation and precipitation. Where the mountain range is high, most of the water vapour will be expended as precipitation on the windward slopes. Once the airmass has crossed the summit, it descends on the leeward slopes. As it descends, the air is compressed and warmed and clouds evaporate. As a result, the leeward slopes receive little moisture, creating an arid or semi-arid environment. For example, Arica, located in the Atacama Desert in Chile and in the lee of the Andes Mountains, receives an average of just 0.5 mm of rain a year.

Figure
1.5   The formation of rain shadow deserts

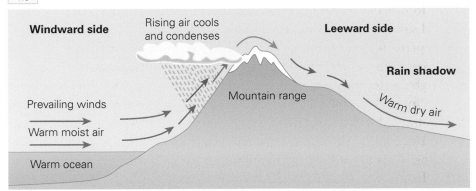

There are many other examples of mountains that act as barriers to rainfall and contribute to arid conditions. They include the Sierra Nevada range in the USA, the Elburz Mountains in Iran, and the Great Dividing Range in Australia.

Activity 3

Using an atlas to help you, draw a sketch map to show the location of the Sierra Nevada mountain range and the arid area in the southwest USA (the Mojave desert and part of the Sonoran desert). Add labels to explain why the Mojave and Sonoran deserts are described as rain shadow deserts.

Cold ocean currents

Several deserts lying along the western coasts of continents owe their formation to cold ocean surface currents. The cold ocean currents are part of the thermo-haline circulation, taking cold water from the poles towards the Equator. The surface currents are associated with the upwelling of cold seawater.

Onshore winds blowing across a cold ocean current close to the shore will rapidly cool in the lower layers (up to 500 m). If the air is chilled by the cold ocean water to its dew point, condensation will take place. The condensation occurring offshore forms mist and fog. Winds carry this mist and fog onshore. This explains the notorious foggy coasts of Oman, Peru and Namibia.

Meanwhile, warmer air aloft creates an inversion that prevents the air from rising and so inhibits cloud formation and precipitation. As the air moves inland it is warmed, lowering its humidity. Mist and fog is soon burnt off by the intense solar radiation, and rarely lasts beyond midday. In these circumstances, there is little likelihood of rain. Figure 1.6 shows the influence of the cold Benguela Current on the Namib Desert.

Figure 1.6 The influence of cold ocean currents on the Namib Desert

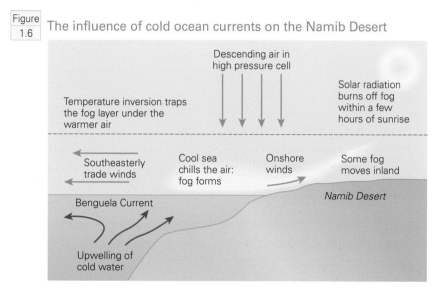

Figure 1.7 Major cold ocean currents

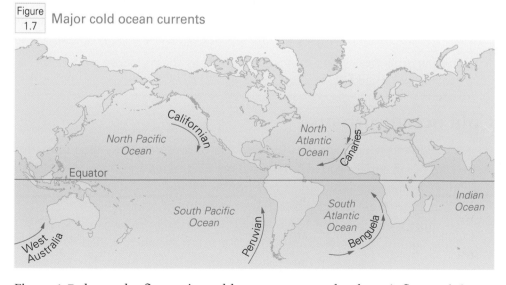

Figure 1.7 shows the five major cold ocean currents that have influenced the formation of coastal deserts:

- the Peruvian or Humboldt Current along the coast of northern Chile and Peru, which influences the Atacama Desert
- the Benguela Current along the coast of Namibia, influencing the Namib Desert
- the Californian Current along the southwest coast of the USA, influencing the Mojave and Sonoran deserts
- the Canaries Current along the Atlantic coast of north Africa, influencing the Sahara Desert
- the West Australia Current along the coast of western Australia, influencing the Western Desert

Continentality

Many arid areas occur within the largest land masses of the world — in dry continental areas. As an airmass moves from an ocean over a continent, it will lose moisture as precipitation, but will not pick up much moisture because little evaporation takes place over the land. Consequently, places with a maritime location will generally have much more rainfall than places inland. Areas in the centre of continents, then, have little rainfall simply because the air has dried out during its long passage over land. This effect can be seen if we look at large continental landmasses such as North America, Asia and Australasia. For example, the Gobi Desert in the Asian interior is so dry because of its distance from the coast — a large part of it is over 2000 km away from the nearest ocean. However, because of the high latitude and high elevation of the Gobi Desert, it is a temperate arid area, and experiences warm summers and cold winters.

Deserts found in the interior of large continents are called continental isolation deserts.

Offshore winds

Many hot arid areas are located where the prevailing wind blows from the land to the sea, thus carrying little moisture. This can be seen in areas such as the Sahara Desert. Despite the proximity to the Atlantic Ocean, the prevailing winds are the northeast trades, and they blow across the large continental expanse of north Africa towards the Atlantic Ocean. By the time these winds reach the Sahara Desert, there is little moisture available for precipitation.

Changes over time

We have seen that one of the most important factors influencing the geographical distribution of arid areas is the general circulation of the atmosphere. However, as a result of global climate shifts, the circulation changes significantly over time. Fluctuations in aridity in areas such as Australia, for example, have generally been associated with the glacial intervals of the last few glacial cycles, with aridity increasing in the inter-glacial periods. As a result of present-day global warming, it is likely that arid and semi-arid environments will become more extensive in the future.

Desertification (the expansion of deserts) is often linked with drought conditions, and the mobility of desert boundaries has been well documented. This is most noticeable in the Sahara, with suggestions that the desert is advancing in a southerly direction at a rate of between 2 km and 5 km a year.

Human influence also has an impact on the extent of arid and semi-arid areas due to desertification. Changes in land use can result in greater demands placed on soil, water and vegetation resources, leading to land degradation, increased aridity and the expansion of deserts. Desertification, its causes and impacts are discussed in Chapter 6.

2 Climate and hydrological processes

The climate of arid environments

Definitions of the arid zone employing climatic indicators such as temperature, rainfall and rainfall effectiveness are discussed in Chapter 1. The climate is a major factor in the formation of landforms and the development of ecosystems.

Köppen's climate classification

There have been many attempts to classify the climates of the Earth; of these the Köppen system is the most widely used. This scientific classification of the climate, first introduced by Köppen in 1900, has been modified several times since, most notably by Geiger in 1961 (see Figure 2.1).

The world is divided into climatic regions that generally coincide with patterns of vegetation and soils. The Köppen system recognises five major climate types, each designated by a capital letter A to E. Group B climates (Table 2.1) are arid, having little rain and a large daily temperature range. They have evapotranspiration rates greater than precipitation rates.

Table 2.1 Group B climates

Category	Description	Example
BWk	Temperate or cold deserts in mid-latitudes. They experience low rainfall and a large temperature range, with warm summers and cold winters. Mean temperatures are below 18°C.	Gobi Desert Takla Makan Desert
BWh	Hot, dry deserts at low latitudes that correspond to the hyper-arid and arid climates defined by UNEP. Mean annual temperature above 18°C and low rainfall, with a winter dry season.	Sahara Desert Central Australia
BSk	Cold, semi-arid or cold-winter deserts found at high latitudes, where mean annual temperatures are below 18°C.	Patagonia Desert
BSh	Semi-arid or steppe areas with a mean annual temperature greater than 18°C. They typically have a short rainy season followed by a dry season.	The Sahel

Figure 2.1 World climate map

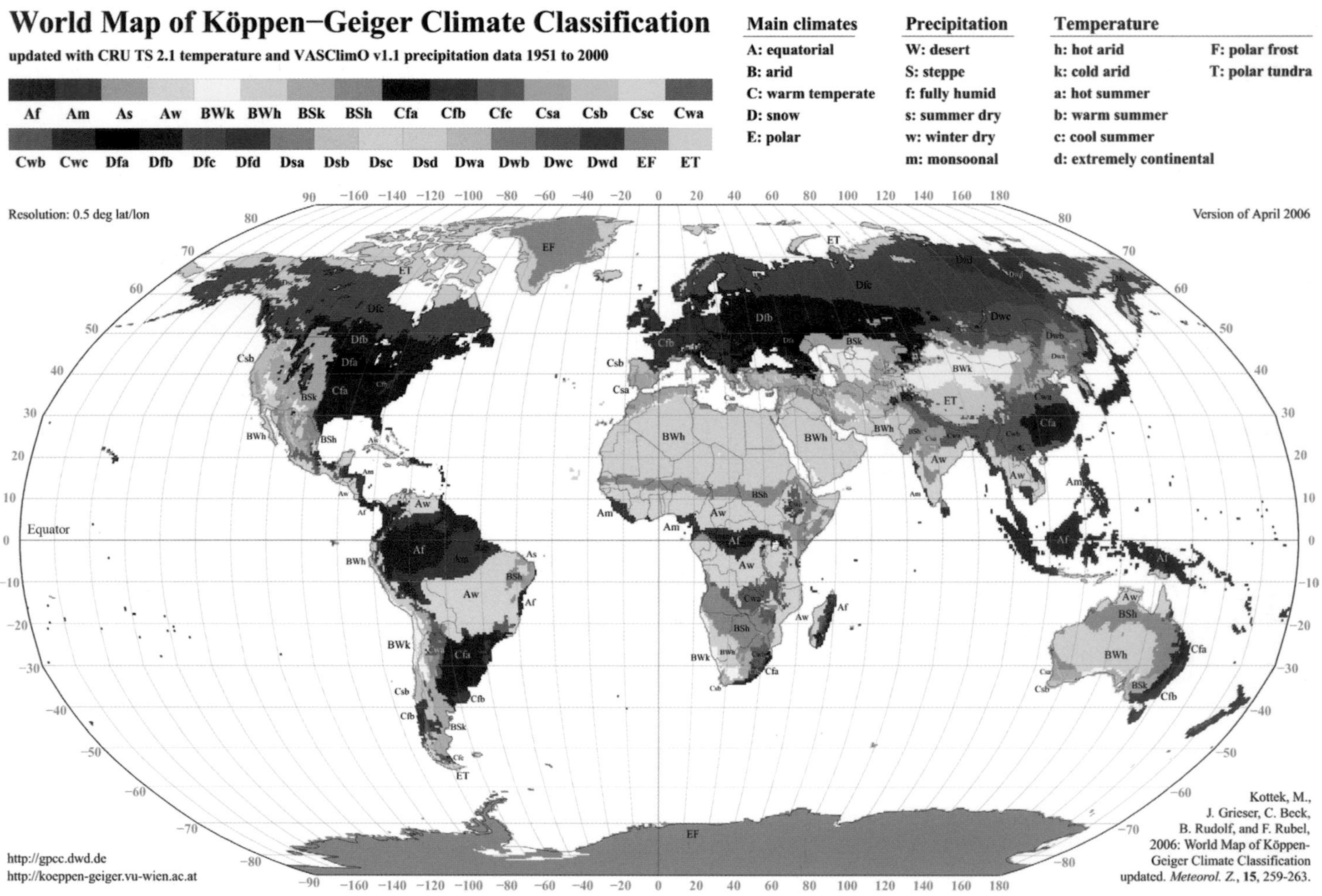
World Map of Köppen–Geiger Climate Classification
updated with CRU TS 2.1 temperature and VASClimO v1.1 precipitation data 1951 to 2000
Af Am As Aw BWk BWh BSk BSh Cfa Cfb Cfc Csa Csb Csc Cwa
Cwb Cwc Dfa Dfb Dfc Dfd Dsa Dsb Dsc Dsd Dwa Dwb Dwc Dwd EF ET
Main climates
A: equatorial
B: arid
C: warm temperate
D: snow
E: polar
Precipitation
W: desert
S: steppe
f: fully humid
s: summer dry
w: winter dry
m: monsoonal
Temperature
h: hot arid
k: cold arid
a: hot summer
b: warm summer
c: cool summer
d: extremely continental
F: polar frost
T: polar tundra
Resolution: 0.5 deg lat/lon
Version of April 2006
Equator
http://gpcc.dwd.de
http://koeppen-geiger.vu-wien.ac.at
Kottek, M., J. Grieser, C. Beck, B. Rudolf, and F. Rubel, 2006: World Map of Köppen-Geiger Climate Classification updated. Meteorol. Z., 15, 259-263.

The two subgroups within this are 'BW' for arid or desert climates, and 'BS' for semi-arid or steppe climates. The 'h' is added as a suffix where mean temperatures are above 18°C, and a 'k' is added if they are below 18°C. A fourth letter can be used to show the seasonality of rainfall: 'w' for winter deficiency, 's' for summer deficiency, and 'd' for year-round lack of rainfall.

Temperature

Temperature regimes

Arid areas are not always hot throughout the year. Annual temperature regimes vary from place to place according to:
- latitude
- altitude
- position within a continent (distance from the sea)
- proximity to cold ocean currents
- albedo

Table 2.2 Classification of deserts according to temperature

Classification of desert	Temperature of warmest month	Temperature of coldest month	Percentage of the world's deserts	Examples
Hot deserts	> 30°C	10–30°C	43%	Central Sahara Arabian Desert
Mild deserts	10–30°C	10–20°C	18%	Southern Sahara Kalahari Desert
Cool deserts	10–30°C	0–10°C	15%	Northern Sahara Mojave Desert
Cold deserts	10–30°C	< 0°C	24%	Gobi Desert Chihuahuan Desert

Influence of latitude

Latitude is the fundamental influence on temperature, as it controls the unequal heating of the Earth's atmosphere. Latitude affects the concentration of solar radiation, the amount of radiation that reaches the ground, and day length.
- Hot arid and semi-arid areas are located at low latitudes. This is because solar radiation is most concentrated in low latitudes due to the sun being at a high angle in the sky for much of the year. The sun's angle changes over the course of the year. On 21 June at noon it is directly overhead at the Tropic of Cancer, while on 21 December it is directly overhead at the Tropic of Capricorn.

- At low latitudes, beams of solar radiation travel through a shorter distance of atmosphere, so that more solar radiation reaches the ground surface. The vertical rays lead to more intense heating of the land. By comparison, in higher latitudes, where the sun's rays slant through the atmosphere at a lower angle, a greater proportion of the solar radiation is absorbed, scattered and reflected. Also the sun's rays are spread out over a larger area of land (Figure 2.2).

- Length of day also affects temperature — the longer the day, the more time there is for the surface of the Earth to absorb insolation. On the Equator, day length hardly varies over the course of the year. Away from the Equator, variation in day length increases with latitude.

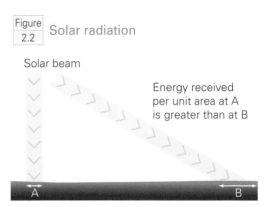

Figure 2.2 Solar radiation

Solar beam

Energy received per unit area at A is greater than at B

A

B

Temperature ranges

Hot deserts have extremes of temperature, both annually (from the hottest to the coldest month) and diurnally (over the course of a day and a night). When comparing places at similar latitudes, it is noticeable that desert areas have a wider range of temperatures than more humid areas. Low latitude and lack of cloud cover result in high levels of insolation. Consequently, temperatures rise rapidly during the day as the intense solar radiation heats up the ground surface. The air above the ground is then warmed by heat transfer through conduction and convection. At night time, temperatures may plummet; ground temperatures sometimes fall below freezing, because clear skies allow terrestrial radiation to escape. The air above the ground is chilled by conduction. An average diurnal temperature range of 15–20 °C is not uncommon in desert areas.

Deserts such as the Atacama in northern Chile and Peru, and the Namib in southwest Africa, which are close to the ocean, experience a lower diurnal temperature range. Temperatures are moderated by the ocean — cool onshore winds lower daytime temperatures that do not reach the extremes of inland deserts. Deserts at high elevations (e.g. the Mojave) also tend to have lower diurnal temperature ranges, and are cooler than those at lower altitudes.

Temperature ranges between summer and winter months are strongly influenced by latitude. In tropical desert areas the mean annual range can be less than 10 °C, whereas in subtropical deserts, with colder winters, the annual range may exceed 25–30 °C.

Temperature extremes

High levels of solar radiation can produce surface temperatures in excess of 80 °C. Air temperatures (recorded in the shade) are lower, though it is not unusual in the hot deserts for air temperature extremes to reach 50 °C and above in the shade. The highest temperatures ever recorded are:

- 57.7 °C in Al-Aziziyah, Libya, on 13 September 1922
- 56.7 °C in Death Valley, California, on 10 July 1913

The **albedo** is the percentage of solar radiation that is reflected by the Earth's surface rather than being absorbed. This has a big influence on surface temperatures, with lighter surfaces reflecting more incoming radiation, which lowers temperatures. For example, white salt crusts, formed where lakes have dried out, reflect approximately 40–80% of the incoming solar radiation. As a result, surface temperatures can be around 10 °C lower on than dark surfaces such as basalt, which absorb a much greater proportion of the solar radiation.

Precipitation

The total annual precipitation in desert areas is low. Annual statistics for rainfall are normally averaged out over a 30-year period. The most hyper-arid areas of the world may experience only a handful of rainfall events over this time. Meigs (1961) noted that hyper-arid lands have at least 12 consecutive months without rainfall.

Seasonal precipitation

In Chapter 1 we saw how the circulation of the atmosphere in the Hadley cell creates an area of permanent high pressure, and therefore an absence of rainfall, at around 30° north and south of the equator. At the equator, the ITCZ is responsible for a belt of heavy convectional rainfall, whose position shifts according to season. As the overhead sun moves over the course of the year between the tropics, there is a corresponding seasonal movement of the Hadley cell and the ITCZ (Figure 2.3).

As the ITCZ moves away from the Equator, it brings summer convectional rainfall to those semi-arid areas located between the hot deserts and the Equator, such as Kano. Although often heavy, this seasonal rainfall is not that effective for plant growth, since it coincides with the period of maximum heat and therefore high evapotranspiration.

Subtropical arid areas on the western and poleward sides of the hot deserts have seasonal rainfall in winter caused by winds converging along the polar front between 50° and 60° north and south of the equator. Along this front,

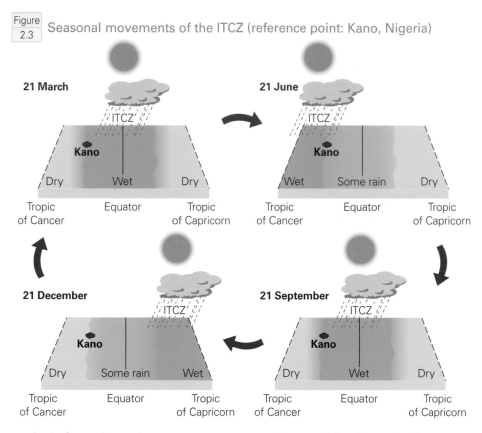

Figure 2.3 Seasonal movements of the ITCZ (reference point: Kano, Nigeria)

cool air from the poles meets warm tropical air and leads to the formation of depressions and rainfall. For example, in southern Arizona during winter, depressions tracking east from the Pacific Ocean bring rainfall to the desert. As the sun tracks south during the northern hemisphere winter, the polar front also undergoes a southerly shift, bringing depressions and rain to parts of northern Africa.

Rainfall in hot deserts is frequently triggered by the intense heating of the ground and convectional updraughts of air, which generate thunderstorms and flash floods. This type of event is more common in summer, since in winter most arid areas experience high pressure.

Rainfall variability

Generalisations about precipitation in desert areas are difficult, as precipitation is unreliable, and varies considerably in terms of frequency, intensity and duration. Hyper-arid areas have the most erratic rainfall, and may experience no rainfall at all for several years — for example, Arica in Chile recorded no rainfall between October 1903 and January 1918. Semi-arid areas, however, have a more clearly-defined seasonal pattern to their rainfall.

Arid regions typically experience wide variations in rainfall both within and between years. This inter-annual variability is summarised in the rainfall variability index:

$$v\,(\%) = \frac{\text{mean deviation from the average}}{\text{the average}} \times 100$$

where

v = temporal variability

Variability is often greater than 30% in arid areas, and may exceed 100% in hyper-arid areas such as the central Sahara Desert. This compares with a typical variability of 10% in the UK. Semi-arid areas have seasonal and more reliable rainfall, with up to 25% inter-annual variability.

Figure 2.4 shows the variation of rainfall experienced in Death Valley between 1887 and 1994.

Figure 2.4 Annual rainfall in Death Valley, 1887–1994

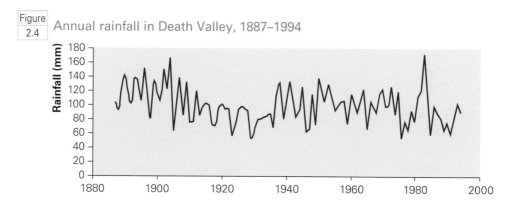

Erratic rainfall patterns and a high inter-annual variability mean that arid areas are often associated with drought. **Drought** occurs when there is a prolonged period with less than average precipitation — this is usually a short-term condition that occurs when the seasonal rains fail. The natural environment can recover from short-term droughts. Nonetheless they can have a devastating impact on the vegetation, livestock, agriculture and people living in the affected area.

Spatial variability of rainfall is also a typical feature of deserts. It is common for only parts of drainage basins to be affected by rainfall and experience runoff at any one time.

Process

Activity 1

Table 2.3 Historic rainfall data for Alice Springs (1949–2008) from the Australian Government Bureau of Meteorology

Year	Annual rainfall (mm)	Year	Annual rainfall (mm)	Year	Annual rainfall (mm)
1949	292.8	1969	184.9	1989	237.6
1950	374.3	1970	175.8	1990	188.8
1951	159.3	1971	125.5	1991	147.8
1952	207.9	1972	271.3	1992	194.8
1953	324.8	1973	449.4	1993	342.4
1954	225.1	1974	782.5	1994	88.8
1955	241.2	1975	601.2	1995	353.4
1956	276.2	1976	551.9	1996	85.2
1957	237.9	1977	372.0	1997	433.4
1958	176.9	1978	428.8	1998	295.2
1959	154.9	1979	308.0	1999	162.8
1960	232.0	1980	169.4	2000	663.8
1961	92.9	1981	309.6	2001	741.2
1962	195.8	1982	293.6	2002	197.8
1963	108.1	1983	548.6	2003	159.4
1964	123.5	1984	289.0	2004	218.8
1965	82.1	1985	120.6	2005	246.4
1966	389.4	1986	394.0	2006	136.4
1967	236.2	1987	182.4	2007	232.2
1968	463.5	1988	375.6	2008	247.4

Source: www.bom.gov.au/climate/data/weather-data.shtml

(a) Calculate the average annual rainfall at Alice Springs between 1949 and 2008.

(b) Calculate the rainfall variability index ($v\%$).

(c) Choose a suitable chart to present the variation in rainfall at Alice Springs.

(d) Repeat (a) to (c) using historic station data for Durham (UK) over the same period. These data can be obtained from the Met Office (**www.metoffice.gov.uk**).

(e) Compare the rainfall variability at Alice Springs and Durham between 1949 and 2008.

Extreme rainfall events

Extreme rainfall events are common in arid areas — more than the yearly average rainfall can fall in just a few hours. It is not unusual for the rainfall total for a single storm to exceed the average annual amount. Rainfall events often occur between long dry spells. Schick (1987) noted that in the hyper-arid areas of the Sahara, a 24-hour rainfall event can be expected to exceed the mean annual rainfall by three to four times once in every 20–30 year period. Extreme rainfall leads to flooding, as a large proportion of the storm water becomes surface runoff. Indeed, a single intense storm is the most effective agent of landscape change in arid areas.

Table 2.4 Extreme desert rainfall events in the twentieth century

Locality	Country	Year	Mean annual precipitation (mm)	Storm precipitation (mm)	Time period for storm precipitation
Chicama	Peru	1925	4	394	24 hours
Lima	Peru	1925	46	1524	24 hours
Swakopmund	Namibia	1934	15	50	24 hours
Tamanrasset	Algeria	1950	27	44	3 hours
Biskra	Algeria	1969	148	210	2 days
El Djem	Tunisia	1969	275	319	3 days

Source: Cooke, Warren and Goudie, 1993

Activity 2

Study Table 2.5, which shows climate data for the Grand Canyon. The weather station is located at 36.04°N 112.10°W, 2068 m above sea level.

Table 2.5 Climate data for the Grand Canyon, southwest USA

Month	Average temperature (°C)	Average maximum temperature (°C)	Average minimum temperature (°C)	Average precipitation (mm)
January	−1.5	5.3	−7.8	36.8
February	0.1	7.2	−6.2	41.4
March	2.8	10.4	−4.0	40.9
April	6.8	15.6	−0.8	24.2
May	11.8	20.8	3.2	17.7
June	17.0	26.9	7.4	12.6
July	20.1	29.0	11.5	48.5
August	18.9	27.4	10.9	57.8
September	15.7	24.2	7.4	37.7
October	9.5	17.9	1.8	30.9
November	3.4	10.8	−3.4	28.8
December	−0.7	6.3	−6.9	40.9

Activity 2 (continued)

(a) Construct a graph to show the climate of the Grand Canyon.

(b) Describe the climate of the Grand Canyon.

(c) Calculate the range of mean annual temperatures and precipitation.

(d) Explain why (i) the precipitation is higher and (ii) the winter temperatures are lower than you might expect to find in an arid area.

Evapotranspiration and potential evapotranspiration

The high levels of insolation and the lack of cloud cover that characterise arid and semi-arid areas mean that evaporation and transpiration levels are potentially very high. Because both rainfall amounts and soil moisture stores are low, actual evaporation is much lower than would occur if water supply were continuously available. Similarly, the sparse vegetation cover means that actual transpiration is low.

Table 2.6 | Definitions for evaporation, transpiration and potential evapotranspiration

Evaporation	Movement of water to the air from stores in the soil, rivers, canopy interception and other water bodies
Transpiration	Loss of water from plants through their stomata
Potential evapotranspiration	The water loss that could occur if there was no limit to the supply of water from the soil

Evapotranspiration can be measured using:

- evaporation pans — shallow metal pans placed on the surface and filled with water, allowing the water loss to be measured
- lysimeters — cylinder tubes placed in the ground around an area of vegetation to measure water loss

Because precipitation in arid and semi-arid areas is so low, irrigation is needed for agriculture. Water for irrigation can be obtained from surface stores — rivers, canals or reservoirs — although high rates of evaporation mean that water is lost quickly from the soil. Irrigation techniques need to be planned carefully so as not to waste precious water reserves.

Rainfall effectiveness

Annual or monthly rainfall figures are not a good measure of water availability in arid areas, since they do not indicate how effective the rainfall will be for plants and crops. Rainfall effectiveness is a measure of the amount of rain that reaches the root zone and becomes available for plant growth. It is determined by both rainfall and evapotranspiration:

rainfall effectiveness = actual precipitation – evapotranspiration

The water budget of an area can be worked out using the following formula:

water budget = $P - PET +/- S$

where
P = precipitation
PET = potential evapotranspiration
S = storage of water

Rainfall effectiveness is influenced by the following:
- Rates of evaporation, which are determined by temperature and wind speed. A large proportion of rainfall is lost to evaporation in hot, dry climates.
- Rates of transpiration. These are determined by the amount and type of vegetation cover.
- Seasonality. Summer rainfall is less effective than winter rainfall because evapotranspiration losses are higher in summer.
- Rainfall intensity. If rain falls in heavy convectional downpours, this will lead to rapid runoff and only a small proportion will infiltrate the soil.
- Soil type. Sandy soils are porous and highly susceptible to drought. Clay soils have a limited capacity to absorb water and cause additional runoff.

Winds

Strong, seasonal winds affect many arid and semi-arid regions. They can be examined at macro-, meso- and micro-scales.

Macro-scale wind systems

Trade winds are the most extensive wind systems in arid and semi-arid areas. They are generated by the movements of air in the Hadley cell (outlined in Chapter 1). From around 30° latitude these winds blow equatorwards, from the northeast in northern hemisphere and from the southeast in the southern hemisphere.

The positioning of the trade winds is linked to the movement of the sun and the Hadley cells and results in trade winds affecting many arid and semi-arid areas for only part of the year. The most powerful and extensive trade winds are the **Harmattan** winds that blow across the southern Sahara Desert and affect west Africa in winter. These winds can carry large quantities of sand and dust, and are important for dune development. The dust is deposited as it reaches the Gulf of Guinea.

Low-pressure conditions in summer, caused by heating of continental interiors, can lead to monsoons. These monsoons are most pronounced over Asia, and result in strong southwesterly winds that blow across parts of Pakistan, India and the Arabian Peninsula. The monsoon winds also transport large amounts of dust from east Africa. The **Shamal** is a northwesterly wind that blows over the Arabian Gulf, carrying dust from Iraq. The **Sirocco** wind is caused by a low-pressure system that develops over the Mediterranean and then moves eastwards. Blowing across north Africa, the warm Sirocco tracks northwards to Europe. It carries large quantities of dust from the Sahara during May and June.

We have seen that, where the prevailing winds over arid areas blow from land to sea, they carry little moisture — for example, the northeast trades blowing across north Africa and the Sahara Desert towards the Atlantic Ocean.

Table 2.7 Examples of seasonal winds affecting the Sahara

Wind	Place	Description
Chelili	Tunisia	A local depression wind that brings hot and dry conditions to northern Tunisia as it blows from the south over the desert area during spring and summer
Djebeli	Tunisia	A wind that descends from the Atlas Mountains, bringing cooler air to Tunisia
Chegui	Tunisia	A wind that blows over the Mediterranean Sea, bringing humid air to places near the coastline. It is common in winter.
Harmattan	Nigeria	A hot, dry wind, associated with dust storms, that blows through the Sahel towards Nigeria; common in November–April
Khamsin	Egypt	A wind that blows through the Nile Valley from south to north during the spring, bringing high temperatures with it. It causes dust storms in Cairo and along the Nile Delta
Sirocco	North Africa	A wind that blows from north Africa to southern Europe and is associated with hot, dry weather and dust storms
Irifi	Western Sahara	A wind that blows from the hotter central and east Sahara to the cooler west Sahara in spring, bringing with it much hotter weather and dust

Meso-scale winds

On a local scale, winds are affected by the topography of an area. Winds deflected around hills and mountains, and funnelled through valleys and mountain passes, increase in strength and consistency. Where mountain ranges obstruct wind flow, winds are forced upwards rather than being deflected around them. The result is strong **anabatic** (upslope) winds and strong **katabatic** (downslope) winds. This effect can be seen around many desert mountain belts, such as the Atlas Mountains in Morocco.

Coastal deserts often experience strong winds because of differences in heating between the land and sea surfaces, which produce localised pressure gradients. These winds are strongest during summer when the temperature gradient is at its maximum. Sea breezes are also common occurrences. The advection fog that is notorious on the coast of Namibia is partly due to local onshore winds.

Micro-scale winds

Wind speeds are greater during the day because solar radiation causes convection, which intensifies the pressure gradient. Wind speeds decrease at night when rapid cooling leads to a stable layer of air near the surface. **Dust devils** are most likely to develop in the middle of the day when solar insolation reaches a maximum. They are thermal vortices, caused by surface heating and a spiralling wind, that carry dust and other debris upwards. The vortex comprises a calm centre of low pressure around which winds rotate at high velocity. Hot air from the surface is carried upwards in a spiralling column, with a compensating downdraught in the centre. Most dust devils move in a random manner associated with light winds. Idso (1974) noted that as many as 50 to 80 dust devils can form in an area per day, and that their heights tended to range from 75 m to 100 m. Their diameters are normally just a few metres.

Sandstorms/dust storms

Sandstorms are common natural hazards in desert areas. They occur when strong winds blow over dry (largely unvegetated) surfaces and pick up loose sand and dust, which is then suspended in the air. The intensive heating of the air causes the lower atmosphere to become unstable, producing stronger winds at the ground surface. Poor farming methods, deforestation and overgrazing all contribute to an increased occurrence of these storms.

Hydrological processes

The hydrological cycle

We have seen that precipitation in arid areas is low and unreliable. A large proportion of the rainfall comes in heavy storms, during which infiltration is limited, overland flow is dominant and evaporation is high. When rainfall is less intense, water percolates downwards and is stored in aquifers. Actual evapotranspiration is roughly equal to rainfall that is stored on the surface, but rates of potential evapotranspiration are much higher. The hydrological cycle can vary on an annual basis since it is controlled by the level and intensity of rainfall, which affects evaporation and infiltration.

In comparison to temperate regions, the hydrology of arid and semi-arid areas is characterised by:

- infrequent, short-lived rainfall events of high intensity
- low interception rates
- rapid rates of evaporation and transpiration
- bare surfaces that limit infiltration
- a greater proportion of overland flow than throughflow
- channel flows that tend to be ephemeral

The water balance

A drainage basin is an open system, characterised by inputs, throughputs and outputs of energy and materials. It is self-regulated by its component variables. Climate is the most important variable influencing fluvial activity in arid areas. Precipitation has a strong effect on vegetation, surface runoff and the sediment yield.

The water balance equation can be used to calculate the inputs and outputs of water in the drainage basin system, taking into account changes in storage in the components of the hydrological cycle:

$$\text{precipitation} = \text{runoff} + \text{evapotranspiration} +/- \text{storage}$$

Arid and semi-arid areas experience a negative water balance due to a combination of low rainfall and high temperatures. This means that annual outputs from evapotranspiration exceed inputs from precipitation. Most locations within the arid zone experience a water deficit for much of the year.

Drainage patterns

While for much of the year little, if any, water lies on the surface of the land, it is evident that water sometimes flows across the surface. Apart from the great **ergs** (sand seas), where infiltration rates are high, dry stream channels are a familiar feature of arid and semi-arid environments.

Infiltration (the percolation of water through the soil) is an important factor in determining surface runoff. There are two basic ways of generating surface runoff or overland flow:

- Hortonian overland flow — where the rate of rain arriving at the soil surface exceeds the rate of infiltration
- saturation excess — where water is prevented from penetrating the soil surface by the rise of the water table in topographic lows and areas adjacent to stream channels, so that groundwater and rainwater both contribute to overland flow

In arid environments, Hortonian overland flow is the more important. Although rainfall events are of low frequency, they are of high intensity. There is a rapid delivery of water to the soil surface, and ponding of water on the surface occurs as the infiltration capacity is exceeded. Saturation excess is almost impossible in arid areas, since the soil is dry when rainfall occurs.

The transfer of water tends to occur in high-magnitude, low-frequency events. Rainfall events may occur years apart, but they have a big impact — even on gentle slopes, overland flow tends to have a high discharge owing to:

- limited interception as a result of lack of vegetation cover
- rainsplash erosion that displaces fine particles and fills pore spaces, which can make the surface impermeable
- shallow soils, allowing little storage of water
- a large proportion of the drainage being **endoreic** — it drains internally and does not reach the sea. Salt lakes are often found in the middle of endoreic drainage systems.

Sources of water

While there is a fundamental lack of water in arid and semi-arid environments, there are several sources that support animal and plant life, and human activities.

Hydrologic and fluvial features

Perennial rivers in arid areas are generally **allogenic** rivers, i.e. they have their origins in wetter environments outside the arid zone. For example, the Colorado River, which flows across the deserts of the southwest USA, derives its water from the snowpack of the Rocky Mountains. Perennial rivers are important to human habitation in desert areas. The Colorado River is heavily managed and controlled through a series of dams, reservoirs, canals, and water-usage quotas. Allogenic rivers also shape fluvial erosional and depositional landforms within desert environments.

Wadis are dry river beds found in the arid and semi-arid regions. Wadi channels are dry for most of the year, and only contain water after rainstorms.

Runoff occurs in flash floods within the wadi, but the water is quickly lost from the channel through a combination of infiltration and high levels of evaporation. The sources for wadis are normally within the arid regions (**endogenic** drainage).

Ephemeral rivers are those that flow seasonally or after storms. They typically have a high discharge and high sediment loads.

Endoreic rivers are those that do not reach the sea. In desert areas, most rivers — whether allogenic or endogenic — drain to inland basins because of high rates of water loss due to evaporation and infiltration. However, some deserts do have areas of **exoreic** drainage, where the rivers meet the sea. For instance, the Loa River in Chile is a perennial allogenic river that flows from the Andes Mountains across the Atacama Desert into the Pacific Ocean. In exoreic basins, rivers transport most sediment to the sea. In contrast, in endoreic systems, sediment accumulates in enclosed basins.

Aquifers (underground layers of water-bearing, permeable rock) occur at various depths. Some aquifers are shallow, and local rainfall events will increase the amount of water within these, but the aquifers in arid areas are more commonly deep underground, such as the Great Artesian Basin in Australia. This is the world's largest deep aquifer, and underlies around three-quarters of the Simpson Desert. The water filling these deep aquifers may have fallen as rain under past climatic conditions. For example, the Great Artesian Basin contains fossil water that is 20000 years old. In the Sahara, there are many fossil aquifers that were filled during the last ice age. While aquifers can supply significant quantities of water to wells, this water cannot be easily or quickly replaced. Overexploitation is therefore a problem: if more water is removed than can be replaced naturally, the resource becomes unsustainable.

Coastal mist and fog

Some coastal arid areas are affected by mist and fog. There is a potential for harnessing this moisture to supplement rainfall for human use. In 1987, Chungungo in Chile became the first place to utilise technology that allowed the harnessing of water from fog to meet its requirements. The fog, known locally as the Camanchaca, regularly drifts inland from the ocean but lacks enough moisture to produce rain. The fog collection system installed on the slopes of the coastal mountains consists of a series of large nylon mesh nets each stretched between two posts (Figure 2.5). They are arranged across the slopes perpendicular to the direction of the landward flow of the camanchaca. Water droplets in the fog condense when they come into contact with the net and drip into a trough below, which feeds a reservoir and a network of pipes.

Chungungo receives just 60 mm of rain a year, yet this system provides an average of 15 000 litres of potable water a day. Today fog collectors provide water to communities in other areas of the world.

| Figure 2.5 | A fog collection system |

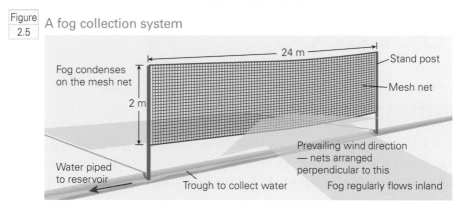

- Fog condenses on the mesh net
- 24 m
- Stand post
- Mesh net
- 2 m
- Water piped to reservoir
- Trough to collect water
- Prevailing wind direction — nets arranged perpendicular to this
- Fog regularly flows inland

Activity 3

Water is fundamental to the development of arid areas. Comment on the sustainability of water sources.

Water problems

Humans living in arid areas are able to utilise many different water sources in order to meet demand, most commonly by damming rivers and wadis. Yet in spite of this, water remains in short supply. An alternative response to water shortage is desalinisation of seawater. This is an expensive process and is used only on a small scale, notably in energy-rich countries such as Saudi Arabia and the United Arab Emirates. There are currently 18 desalinisation plants along the Red Sea coast of Saudi Arabia.

The lack of water in arid areas can lead to transboundary conflicts over water supplies. Particular problems occur where the same water source, such as a river, is shared by several countries — for example, the River Nile in Africa and the River Jordan in the Middle East. Agriculture takes the most water, but industry and domestic use also make high demands. If too much water is removed upstream, it reduces water availability to places downstream.

Activity 4

Using the internet, investigate the potential transboundary conflicts over water supplies from either the River Nile or the River Jordan. Comment on the river management scheme.

3 Geomorphological processes

Weathering processes

Weathering is the process of breaking down a rock (in situ). It is the funda-mental mechanism leading to rock disintegration and the production of sediment. Rates of weathering in hot arid and semi-arid environments are influenced by the climate, rock material properties, and the stress–strain behaviour of materials exposed to weathering. The results of weathering are four-fold:

- major landforms
- small-scale geomorphological forms at the ground surface
- granular disintegration of rock and the production of sediment for mobilisa-tion by other processes
- soil development

Weathering processes produce **regolith**, which is an unconsolidated layer of material found above the solid rock. This material can be eroded and trans-ported easily, and may be involved in eroding other materials.

In arid areas, weathering comprises mainly physical and chemical weathering types, although biological weathering also has some influence. Physical weathering is the breakdown of rock by mechanical forces. Chemical weathering is the breakdown of rock through alterations to its chemical compo-sition. Across the arid zone, insolation, frost and moisture all induce rock disintegration. Of particular importance is the presence of natural salts, which can have both physical and chemical consequences, such as crystal growth from solution, hydration and thermal effects. In arid areas, chemical weathering is generally slow and physical weathering more rapid.

Physical weathering

Physical weathering is the main weathering process leading to the breakdown of rocks in arid and semi-arid areas. This mechanical breakdown of the rocks can be rapid, since the absence of plant cover or well-developed soils means that bedrock is exposed over large areas. There are three main physical weathering processes: insolation weathering, salt weathering and freeze–thaw weathering.

Insolation weathering

While air temperatures can reach over 40°C, surface temperatures get even higher and can exceed 80°C during the day in the summer months. Surface temperatures can then drop to near freezing at night. This means that the rock surfaces will expand and contract daily. The mechanical fracture and breakdown of the rock caused by heating and cooling is known as **insolation weathering**. This can occur in a number of different ways, determined by rock type, structure, chemical composition and colour.

Granular disintegration occurs as a result of the large temperature range, which causes the minerals within the rock to expand and contract at different rates. The light and dark minerals within rocks (e.g. granite, which contains both black mica and white quartz crystals) heat and cool at different rates, which leads to stresses within the rock, and eventual disintegration.

Exfoliation is the peeling of surface rock layers caused by insolation weathering. High temperatures during the day mean that surface rock layers, which are more exposed to the high temperatures, heat up and expand. At night these layers then cool and contract. The repeated expansion and contraction of the surface rock layer eventually cause it to peel or flake off. The process is referred to as 'thermal exfoliation' or 'onion-skin weathering', since sheets of the rock fall away like the concentric layers of an onion (Figure 3.1)

Block disintegration is caused by the repeated heating and cooling of rocks that are well jointed, such as limestone. The rocks break down along the joints and bedding planes, since these are the main lines of weakness.

| Figure 3.1 | Thermal exfoliation: Death Valley, California |

Lucy Cole

Previous thinking was that insolation weathering was the main process leading to the breakdown of rocks in arid environments, and that this could occur without the presence of moisture. However, laboratory tests by

Griggs (1936) proved that the temperature ranges in deserts are not sufficient on their own to cause the breakdown of rock, and that the presence of water from dew or fog acts as the trigger. Many scientists now believe that there needs to be at least some moisture present in order for insolation weathering to occur.

Salt weathering/crystal growth

Salt weathering occurs when the salt in the rocks crystallises out of solution. The high temperatures draw saline groundwater to the surface. Evaporation of the water on the surface leaves behind the salt crystals. The growth of salt crystals between pores and joints in the rock creates stresses in the rock, causing it to disintegrate — this can lead to either granular disintegration or block disintegration. For example, crystals of sodium sulfate can expand by 300% in areas of high insolation. This is a major cause of weathering in desert areas, particularly in porous, sedimentary rocks like sandstone.

Salt weathering is more important in desert environments than in more humid environments: where moisture is available, salts are dissolved by rainwater and removed in solution by streams and rivers. However, in the drier desert environments salts such as sodium chloride are not removed and therefore accumulate in inland drainage basins.

Freeze–thaw weathering

Water trapped in confined rock joints and crevices expands by 9% of its volume when it freezes. The pressure exerted in the rock by the ice is enough to cause it to shatter. The shattered rock forms scree.

Freeze–thaw weathering is only possible in areas where temperatures fluctuate above and below freezing, and there is sufficient moisture available. The process is more rapid if temperatures fluctuate frequently across the freezing threshold. These conditions are more likely to be found in the semi-arid, mid-latitude environments or in areas of higher altitude such as mountainous regions. Many subtropical deserts are at altitude and have seasonal temperatures that fall below freezing. Frost is common in the Namib Desert and in the deserts of the southwest USA.

Activity 1

(a) Draw a graph to represent the data in Table 3.1.
(b) What evidence suggests that freeze–thaw weathering is active? Name the months of the year in which you would expect freeze–thaw weathering to occur and give reasons for your choices.

Activity 1 (continued)

Table 3.1 Average maximum and minimum monthly temperatures at Bryce Canyon, Utah, recorded at the weather station 2412 m above sea level

Month	Average maximum temperature (°C)	Average minimum temperature (°C)
January	1.6	−13.4
February	3.1	−11.6
March	5.6	−8.6
April	10.7	−4.9
May	16.6	−0.7
June	22.7	3.7
July	26.0	7.7
August	24.3	6.9
September	20.2	2.4
October	14.5	−2.8
November	6.5	−7.8
December	2.0	−12.7

Chemical weathering

An important influence on the rate of weathering is the availability of water; therefore the chemical breakdown of rock in the arid zone is slow. It was once thought that chemical processes were rare in the arid areas, but today their importance is widely recognised. While the defining feature of such areas is the scarcity of water, few deserts are completely dry. Occasional downpours mean that water enters the pores and joints in the rocks, allowing chemical action to weaken the structure. Dew and fog that occur in arid areas as the air is chilled at night also add water to the ground surface. The main chemical weathering processes operating in arid areas are wetting and drying, hydration, oxidation and solution.

Wetting and drying

Water in arid areas may come from flash floods, seasonal rains, nocturnal dew or coastal fog. If water saturates the rock, this can encourage the clay minerals within the rock to swell. As the water evaporates, the rock contracts again as it dries. Repeated expansion and contraction of the rock causes it to disintegrate.

Hydration

Hydration, which is the absorption of water by minerals, is widespread. When minerals such as salts within the rock absorb water, they swell and expand. The increase in volume causes stress within the rock, and zones of weakness develop, which leads to granular disintegration and the flaking of rocks exposed near the ground surface.

The chemical addition of water molecules to minerals can produce new mineral compounds — for example, water added to the mineral anhydrite produces gypsum.

Oxidation

Oxygen dissolved in water reacts with some minerals to create oxides and hydroxides. This can lead to a red staining on rocks such as sandstone. The minerals that are oxidised also increase in volume, which weakens the rock.

Solution

Some minerals can be dissolved in water, particularly if the rainwater is slightly acidic. For example, sandstone, which contains calcite, can be affected; calcite can leach from the rock, again resulting in the flaking of surface rock layers.

Biological weathering

Biological weathering is limited in arid environments owing to the sparse vegetation cover, but it does have some influence on rock disintegration. Some of the plants found in arid areas have long root systems that enable them to reach down to the groundwater. The roots that grow in the cracks in the rock can widen the cracks as they grow.

Lichen and algae found on the surface of rocks can cause micro-morphological changes to the rock. Respiration releases carbon dioxide that can then be used for chemical processes such as carbonation.

Aeolian processes

Aeolian erosion

The three main processes of **aeolian** (wind) erosion are **deflation**, **corrasion** and **attrition**. The amount and rate of aeolian erosion is determined by several factors, as shown in Figure 3.2.

Figure
3.2
Factors that affect the amount of aeolian erosion

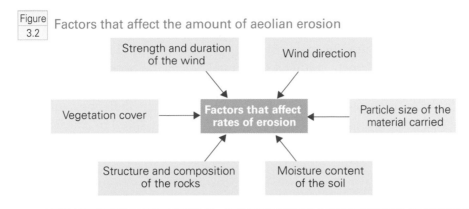

Activity 2

Explain how the factors shown in Figure 3.2 are likely to affect the amount and rate of aeolian erosion in arid and semi-arid areas.

Deflation

Deflation is the dominant process of aeolian erosion in arid landscapes. The loose and fine regolith is removed by the wind (Figure 3.3). The process is active in arid areas owing to the sparse covering of the land surface by either vegetation or soil. The dry conditions also mean that the loose sand and dust can easily be eroded. Deflation operates through the progressive removal of small material until only the larger material is left behind, forming a stony desert surface.

Figure
3.3
The selective removal of finer particles by deflation

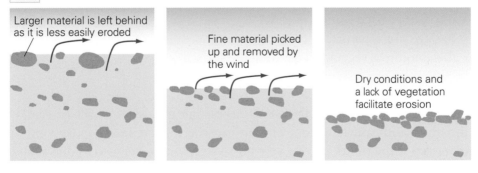

Corrasion

Wind corrasion is erosion caused by the abrasive action of wind-borne particles driven against rocks. The particles act like sandpaper. Over time, as sand and smaller-grained particles are driven against the rock by the wind, they carve the rock into different shapes and features. It is the larger particles that are more erosive. Because the grains of sand are heavy, the sand-blasting effect of wind abrasion is largely confined to a couple of metres above the surface of the ground.

Wind corrasion takes place much more slowly than deflation, and may take 100 years to erode a layer of rock 1 mm thick. The speed of abrasion will depend on the strength of the rock and the velocity of the wind.

Attrition

Attrition takes place as grains of sand carried by the wind collide with each other and become smaller and rounder as they do so.

Aeolian transportation

Transportation of sand and other fine particles occurs when the wind velocity exceeds the critical threshold needed. Near to the surface, although wind speeds are reduced, the amount of turbulence increases. The movement of the sediment is induced by forces of drag and lift. Drag occurs as a result of differences in pressure on the windward and leeward sides of the sand grains.

Processes of transportation of sand (particles between 0.05 mm and 2 mm in diameter) include **suspension**, **saltation** and **surface creep**. While all three processes operate in arid areas, the movement of particles will be influenced by:
- wind speed and duration
- wind direction
- the degree of turbulence
- the nature of the surface material
- the size of the particle
- the amount of vegetation covering the surface

Suspension

Suspension occurs where the wind carries fine dust such as silt or clay, or the finer sand particles. It operates on particles that are less than 0.15 mm in diameter. If the winds are of a very high velocity, this can lead to dust clouds and sandstorms, which are fairly frequent occurrences in desert areas (Figure 3.4). The smaller particles can be transported over long distances, and may even be carried beyond the area of desert. For example, sand from the Sahara can be carried as far as the UK, where it falls as 'red rain'.

Figure 3.4 A sandstorm in the Sahara Desert

Lucy Cole

Saltation

Particles measuring 0.15–0.25 mm in diameter move in a series of small leaps along the ground (Figure 3.5). This is the main form of wind transportation in desert areas. The grains of sand are bounced along close to the ground surface during periods of strong winds. Owing to the weight of the particles, this process occurs within a couple of metres of the ground surface, with most of the movement close to the ground. A gust of wind will pick up grains of sediment, move them forward, and then drop them. As the falling grains hit other grains of sand, they set the stationary grains in motion.

Creep

Heavier grains of sand (above 0.25 mm in diameter) are rolled or pushed along the ground by surface creep (Figure 3.6). They are too heavy to be lifted and so will either be rolled along the surface by the wind, or impacted by other wind-blown particles that push them forward. Larger grains may be dislodged by the particles moved by saltation.

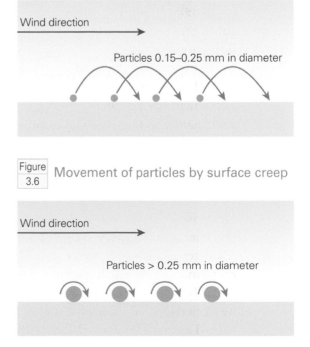

Figure 3.5 Movement of particles by saltation

Wind direction

Particles 0.15–0.25 mm in diameter

Figure 3.6 Movement of particles by surface creep

Wind direction

Particles > 0.25 mm in diameter

Activity 3

(a) Using the data in Table 3.2, plot the entrainment velocity curve for particles of different size (**entrainment velocity** is the velocity at which a particle will be lifted). The particle diameter should be converted to a logarithmic scale. Particle diameter will appear on the *x*-axis; velocity on the *y*-axis.

(b) Describe and explain the relationship between the **threshold velocity** (the point at which grains begin to move) and particle size (hint: consider the effects of gravity and cohesion on the particles).

Table 3.2	Wind erosion and transport: the relationship between particle size and threshold velocity

Particle diameter (mm)	Threshold velocity (cm s^{-1})
0.01	35.4
0.02	28.1
0.04	20.9
0.07	15.9
0.1	17.2
0.2	22.8
0.4	27.2
1.0	46.3
2.0	53.1
3.0	63.2

Aeolian deposition

As the velocity of the wind decreases, it will reach the point where it can no longer transport the particles that it is carrying, and so sediment will be deposited. The sediment carried in suspension will fall and build up on the ground surface. It may be that obstacles in the path of the wind slow down the wind velocity on the downwind (leeward) side. Figure 3.7 shows how vegetation in the path of the wind can lead to a build-up of sand behind the plant.

Figure 3.7 Sand shadows formed on the leeward side of desert vegetation

Lucy Cole

Fluvial processes

Rainfall in arid areas may be limited, but runoff is nevertheless responsible for creating many desert landforms. Erosion by water is mainly limited to the action of streams and rivers that flow during and after periods of heavy rain. These streams can be so powerful that they cause flash floods. This is because a large proportion of the water following rainfall events in arid areas becomes surface runoff. This is due to:

- heavy seasonal and convective rainfall
- limited vegetation cover to intercept rainfall
- impermeable soil crusts, baked hard by the high temperatures

After a torrential downpour, water initially flows as **sheet runoff**, spreading across the ground surface before running into either permanent channels or temporary channels called **wadis**.

The discharge in the channels is initially high, causing a lot of erosion to occur, abrading boulders and scouring the channel. Processes of fluvial erosion include corrasion (abrasion and hydraulic action), attrition and corrosion (Table 3.3).

Table 3.3 Processes of fluvial erosion

Abrasion	Coarse sediments carried by the water grind against and wear away the bed and banks of the channel. The transport of coarse bedload scours the channel.
Hydraulic action	The force of the running water can erode the rock, particularly where the banks are made of unconsolidated material. The pressure of the water running into the joints and bedding planes of the rock further weakens the rock, eventually leading to the removal of blocks of rock.
Attrition	As the bedload is transported, it scrapes along the channel bed and collides with other sediment. It is gradually worn down, sharp edges are removed and the material becomes smaller and rounder.
Corrosion	The rocks can be broken down by the chemical action of water — slightly acidic water dissolves carbonate rocks such as limestone.

The Hjulström curve (Figure 3.8) shows the velocities required for the entrainment and movement of particles. It also indicates the point at which transportation and deposition will take place, according to particle size and velocity.

High discharge and the ready availability of unconsolidated dry sediment allows for the transport of large amounts of sediment. Thus streams and rivers in desert areas tend to transport high sediment loads. Large rock particles can be carried, but because discharge declines rapidly downstream and rivers quickly lose competence, large sediments are unlikely to be carried long distances or for extended periods of time.

 The Hjulström curve: sediment movement and energy

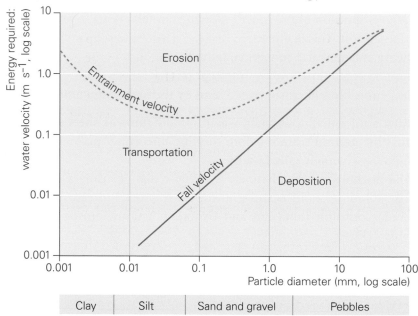

| Clay | Silt | Sand and gravel | Pebbles |

Activity 4

(a) Make a copy of Figure 3.8. Place the following statements in the most appropriate place on the graph:

(i) Particles of 1 mm will start to be lifted at this velocity

(ii) Pebbles will be transported at this velocity

(iii) The velocity is not high enough to lift clay particles from the bank

(iv) As velocity falls, sand will start to be deposited at this point

(v) Erosion of all particle sizes will take place at this velocity

(vi) The velocity is too low to transport any particles larger than clay and silt

(b) Investigate reasons why high velocities are required to erode both the finest and the coarsest particles.

4 Arid landscapes and landforms

The landscapes of both arid and semi-arid areas are shaped by a combination of wind and water, though the relative importance of the two agents has been a matter of debate for geomorphologists. Until the 1960s, it was thought that wind action was the dominant process in arid environments. However, by the 1970s and 1980s this view had changed. Today it is argued that water action is dominant, particularly when explaining the relict landforms from past climatic conditions. Even so, some geomorphologists argue that the importance of water action has been over-emphasised. The debate continues, and it is clear that, whatever the final conclusion, both wind and water have played funda-mental parts in shaping the landscapes and landforms of arid and semi-arid environments.

The wide range of landscapes and landforms that develop in arid and semi-arid areas are determined by several factors:

- the underlying geological structure
- processes of erosion
- wind action
- water action

Arid areas often contain landforms that look similar but that have formed in distinct and different ways. This is the concept of **equifinality**. However, more research is needed in some areas to explain fully how some landscapes and landforms have developed.

Arid landscapes

The term 'arid landscape' refers to the large geological surfaces over which the arid zone extends. They may cover areas of thousands of square kilometres. Arid landscapes are characterised by abrupt changes in slope and by angular surfaces — which contrast sharply with the rounded landscapes found in more humid areas. This is because the sparse vegetation cover exposes the geological structures, and because vegetation and plant root systems don't interrupt geomorphological processes. Within the landscape, there will be a large diversity of landforms that vary according to local circumstances.

At the broadest level, Mabbutt (1977) divided the world's arid lands into two geomorphological types:

- shield drylands — tectonically stable areas consisting mainly of plains or eroded surfaces, for instance the Arabian Peninsula and Australia
- basin-range drylands — tectonically active areas such as steep-land environments, found for instance in the South American deserts and Iran

This division is based on major differences in geological structure and relief, with the idea that these differences will produce different physiological responses to aridity. Many arid areas show striking contrasts and contain a variety of landforms. For instance, sand dunes occupy less than 1% of the American deserts but almost 30% of the Sahara. Mabbutt separated desert landscapes according to geology, landforms and surface materials. He identified seven major landscape units: uplands, shield plains, stony deserts, karst plains, riverine deserts, sand deserts and desert lakes.

The processes that shape the landscape change as the climates of arid areas fluctuate over long timescales. Therefore, many arid landforms and landscapes are the product of past processes of weathering and erosion that are no longer active. For example, freeze–thaw and hydration were more significant in wetter, colder climates. Arid landscapes and landforms are not static; they are dynamic systems constantly undergoing change.

Weathering landforms

Weathering is the fundamental process that leads to the breakdown of rocks and plays a key role in the development of landforms and the morphology of desert surfaces (as outlined in Chapter 3). A wide range of landforms owe their formation to weathering processes. They include desert varnish, alveoles and tafoni.

Desert varnish

Desert varnish consists of a thin, shiny, dark red or black layer on exposed rocks and rock surfaces in arid regions. Desert varnish will only occur on stable rock surfaces that are not exposed to frequent rainfall or aeolian abrasion, and can often be seen coating canyon walls (Figure 4.1). Since it takes thousands of years for a coat of desert varnish to form, it is rarely found on surfaces that can be easily eroded.

Lucy Cole

Figure 4.1 Petroglyphs carved in desert varnish, Glen Canyon, USA

Desert varnish is composed of clay minerals, oxides and hydroxides of manganese and iron, and other particles such as sand grains and trace elements. The main elements are the manganese and the iron, with the proportions of these elements influencing the colour:

- If the desert varnish is rich in manganese, it will be black.
- If the desert varnish is rich in iron, it will be a more red/orange colour.

The longer a rock is exposed, the darker the desert varnish will become. The varnish may be chipped or flaked where rockfalls and other stones strike it.

The origin of desert varnish is complex, and it is not fully understood. It was originally thought to develop when iron and magnesium oxides evaporated up through the rock, leaving a deposit behind. However, it is now believed that the coating comes from sources from without, rather than within, the rock. A large proportion of desert varnish is made up of fine, wind-blown particles. The clay then catches the iron and magnesium oxides that are created by chemical reactions between other substances in the high desert temperatures. A third theory is that the varnish is formed by colonies of microscopic bacteria that live on the rock and absorb trace quantities of iron and magnesium from the atmosphere, and then oxidise it and emplace it as layers of oxides on the rock. Microorganisms live on most rock surfaces, since they are able to feed from both organic and inorganic sources.

Alveoles

Alveoles are small hollows or cavities found on the surface of rocks (Figure 4.2). They are caused by a combination of crystal growth, wetting and drying, and hydration; their development may also be helped by wind abrasion. The hollows range from about 5 cm to 50 cm in diameter, and occur in clusters, appearing as a honeycomb pattern. Thin walls separate the hollows, and these partition walls become strengthened by case-hardening. **Case-hardening** is a hard layer of salt-encrusted rock. In times of high rainfall, water soaks into rocks such as sandstone and mixes with the calcium. During hot, dry periods, water evaporates from the rock. As water evaporates, some of the dissolved calcium is also drawn to the surface by capillary action and forms a hard layer. The remaining calcium is not distributed evenly within the rock, leaving some areas that are low in calcium and softer than the case-hardened, high-calcium areas. As the rock erodes, the softer parts of the rock experience a quicker rate of erosion than the harder areas, forming a series of hollows and cavities.

 Alveoles in southwest USA

Lucy Cole

Tafoni

Tafoni are rounded, natural rock cavities or caves that form in a similar way to alveoles on rock surfaces. The tafoni hollows can be several metres in diameter. They have arch-shaped entrances, and may form along lines of weakness in rocks such as joints and bedding planes. They commonly occur in granular rock such as sandstone. Tafoni tend to develop in clusters, normally on vertical or inclined surfaces.

Aeolian landforms

Wind erosion through deflation and corrasion (abrasion) picks up loose sand and dust, modifying arid landscapes and creating new landforms.

Features created by deflation

Deflation occurs when wind picks up and removes loose, unconsolidated material. This creates deflation hollows, salt lakes and salt pans, and desert pavements. Deflation is the dominant erosional process in desert areas.

Deflation hollows

Deflation hollows are large, enclosed, surface depressions that have been created by the removal of large amounts of loose particles by the wind (Figure 4.3). Deflation hollows can range from shallow features just a few metres across, to much larger landforms several kilometres long and tens of metres deep. As closed basins, deflation hollows collect runoff. The shape of the hollows varies: they can be irregular or symmetrical. The world's largest deflation hollow is the Qattara Depression in the western Egyptian desert, which reaches a depth of over 100 m below sea level. Millions of tonnes of sand and other material have been removed from this hollow by the wind.

Deflation hollows can be eroded downwards until they reach the water table and form **salt lakes** and **salt pans**.

Figure 4.3   The formation of a deflation hollow

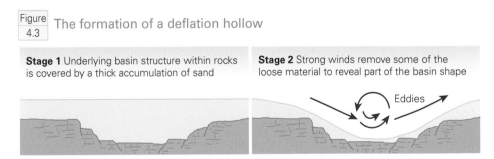

Stage 1 Underlying basin structure within rocks is covered by a thick accumulation of sand

Stage 2 Strong winds remove some of the loose material to reveal part of the basin shape

Eddies

Desert pavements

Desert pavements are surfaces of stones, tightly packed together, that rest on a finer material such as sand, silt or clay. This is the most common type of desert surface. The surface of the desert pavement may be coated with desert varnish.

There are two main theories of formation. The most commonly held theory suggests that desert pavements are formed by deflation as the wind selectively removes the finer material (Figure 3.3, page 39). Clays are removed first, followed by silt and, finally, sand. Sometimes the wind will also be strong enough to move small pebbles. The larger, coarser particles that are too big for the wind to pick up are left behind, accumulating as **lag deposits** on the surface. Eventually the surface is almost entirely formed of the coarse, closely-packed stones that protect the finer particles underneath from erosion. This surface is now in equilibrium.

A second theory proposes that processes such as frost heave, salt heave and cycles of wetting and drying move stones to the top of the soil surface over time.

Alternative names for desert pavements include 'gibber plains' in Australia, 'gobi' in Mongolia and 'reg' in the Arabian and Sahara deserts.

Activity 1

Using your knowledge and understanding of desert geomorphology and climate, evaluate the strengths and weaknesses of the two theories of desert pavement formation.

Landforms of aeolian corrasion

Corrasion occurs when sand that is carried in the wind abrades rock surfaces. This creates landforms such as sculptured rocks and ventifacts. Less resistant rocks such as sandstone are easily eroded by aeolian corrasion.

Ventifacts

A ventifact is a cobble or pebble that has been shaped by the wind-blown sand. Ventifacts have smooth sides separated by sharp edges and are normally just a few centimetres in size. The side facing the prevailing wind will be abraded and faceted (Figure 4.4). The lee side of the facet is protected from the wind by a sharp edge.

Ventifacts with classic, clearly defined facets are described as **einkanter** (one-faceted), **zweikanter** (two-faceted) and **dreikanter** (three-faceted). A dreikanter,

| Figure 4.4 | Formation of ventifacts |

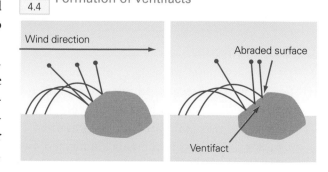

Wind direction

Abraded surface

Ventifact

with three ridges, is pyramidal in shape (Figure 4.5). Ventifacts may form multiple facets either where the rock has been moved by the wind, or where the prevailing winds change direction according to season.

Rock pedestals and zeugens

We have seen that wind corrasion in desert areas is concentrated up to 1.5 m above the ground surface. It follows that undercutting of rocks by wind-blown particles is commonplace in arid environments. **Rock pedestals** are mushroom-shaped rocks that occur where an isolated rock has been eroded more rapidly at the base, causing it to have a top-heavy shape (Figure 4.6).

A similar process can create a narrow ridge of land rather than an individual block, with a protective cap rock above it — a feature called a **zeugen**. Zeugens may owe their formation not just to the increased corrasion at lower levels, but also to lithology: where rocks are in horizontal layers, softer layers are more easily eroded, leaving the more resistant layers behind (Figure 4.7).

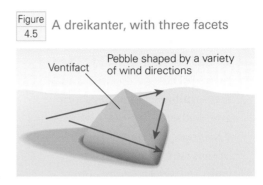

Figure 4.5 A dreikanter, with three facets

Ventifact — Pebble shaped by a variety of wind directions

Figure 4.6 Pedestal rock in southwest USA

Lucy Cole

Figure 4.7 The formation of a zeugen

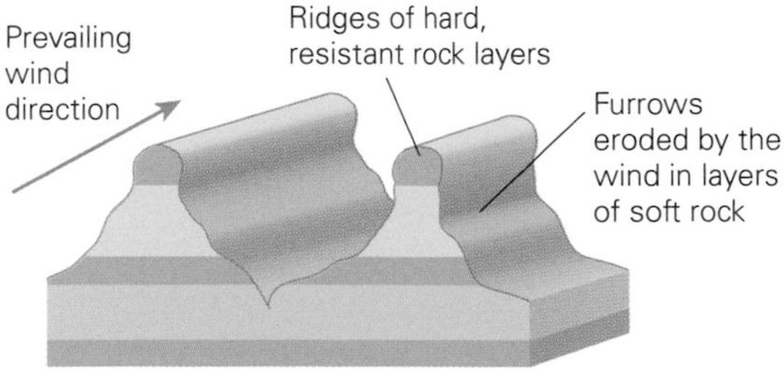

Prevailing wind direction

Ridges of hard, resistant rock layers

Furrows eroded by the wind in layers of soft rock

Yardangs

Sculptured rocks created by wind erosion include **yardangs** — streamlined, steep-crested, linear ridges of clay, silt or rock. They are sculpted predominantly by abrasion, although deflation also contributes to their formation. Yardangs consist of a series of parallel rills or troughs, separated by regular ridges that may be undercut at the base. They form in areas where the rock strata are aligned vertically. The softer areas of rock are eroded, leaving behind the harder areas (Figures 4.8 and 4.9). Yardangs form parallel to the direction of the prevailing wind. The ridges occur in groups, and the troughs between them are scoured by the wind. Yardangs can reach lengths of several kilometres and can vary in height from a few centimetres to over 100 m. The end facing the wind is the highest and widest, and the features are normally about four times longer than they are tall. In 1905, Sven Hedin studied the yardangs in the Turkmenistan desert, and concluded that a wind rill of 6 m would take around 1600 years to form. That amounts to an annual rate of erosion of 4 mm.

Figure 4.8 Plan and profile of a yardang

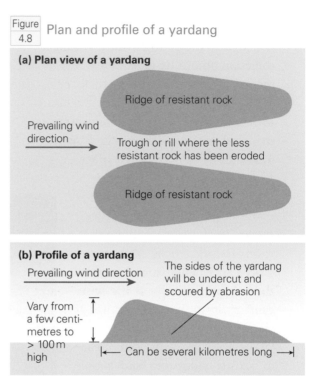

(a) **Plan view of a yardang**

Ridge of resistant rock

Prevailing wind direction

Trough or rill where the less resistant rock has been eroded

Ridge of resistant rock

(b) **Profile of a yardang**

Prevailing wind direction

The sides of the yardang will be undercut and scoured by abrasion

Vary from a few centimetres to > 100 m high

Can be several kilometres long

Figure 4.9 Yardangs in the Gobi Desert, China

George Steinmetz/SPL

Aeolian transportation and deposition

Sand

Sand consists mainly of particles of quartz and feldspar, and is formed by the weathering of rocks such as sandstone and granite that are rich in these minerals. Winds around desert regions pick up sand from riverine environments and transport it long distances before it is deposited in the desert. The deposition and build-up of sand occurs where the wind velocity is reduced — this may be the result of an obstacle in the path of the wind such as a rock or a plant (where wind will be slowed on the downwind side of the obstacle) or a hollow or depression on the land surface. Sand may also accumulate in areas where there is an increase in surface roughness, which also reduces wind velocity.

The proportion of sandy areas varies in each desert — 30% of the Sahara Desert has sand covering the surface, compared to just 1% in the deserts of the USA. Overall, around 20% of the hot arid and semi-arid zones are covered with sand. Large sandy areas without significant dune development are called **sand sheets** — an example is the Selima Sand Sheet in Sudan. These areas without sand dunes tend to be formed from coarser sand, which is less likely to be moved by the wind. However, much of the sand in arid environments accumulates in extensive sandy areas called **sand seas** — known as 'ergs' in the Sahara — which contain a variety of landforms of varying size and scale, including numerous types of dune. Large sand seas occur in Australia, north Africa and Asia.

Figure 4.10 Ripples on a sand dune in the Sahara Desert

Lucy Cole

Ripples

Sand ripples are small-scale features consisting of a series of regularly spaced crests and troughs that have formed at right angles to the wind (Figure 4.10). They form as a result of both surface creep and saltation. Sand is eroded from the windward side of the crest, and then deposited on the leeward side; so ripples are transportational features that migrate downwind. The height of the crest and the distance between crests is determined by the wind speed and the coarseness of the sand particles. Crest heights range from a couple of millimetres to 50 cm. Some degree of sediment sorting can be observed, with the coarsest sand found on the crests and finer sand in the troughs. The spacing of ripples is related to the average distance that the grains of sand jump during saltation.

Sand dunes

Sand dunes result from the build-up of sand blown into mounds and ridges by the wind. Dunes are the main depositional feature of arid areas. In order for sand dunes to develop, there needs to be a sufficient supply of sand, wind velocities that are strong enough to transport it, and topographic conditions that encourage sand particles to accumulate. Dunes normally (but not always) reach heights of up to 30 m, since as they increase in height, wind speed also increases, until erosion becomes as important a process as deposition.

While dunes are commonly regarded as depositional features, they may also (perhaps more accurately) be seen as transportational features — they move in the same general direction as the prevailing wind. Material is moved by saltation up the windward side, and then falls down the steep leeward side through a combination of flows, slides and slumps. The speed of movement of the dunes is determined by wind speed, prevailing wind direction and the size of the sand grains. The sand forming dunes is sorted: as the wind blows across the surface it causes the grains to shake. This process has a natural sieving effect, causing the smaller sand grains to sink below the larger grains, which stay on the surface.

The shape and size of dunes will depend on:
- the availability and amount of sand
- wind strength, direction and duration
- the amount and type of vegetation
- the nature of the ground surface

Dune heights can vary from 300 m, in the case of giant dunes in areas such as the Namib and Gobi deserts, to just a few millimetres. Dune shape depends not only on the direction of the prevailing wind but also on whether the main winds come from one or more directions. Where there is more than one wind direction, the shape of the dune becomes more complex. Figure 4.11 shows the main dune shapes.

Where the dominant wind direction changes seasonally, dunes will respond by adjusting their form. This happens in the deserts found between latitudes

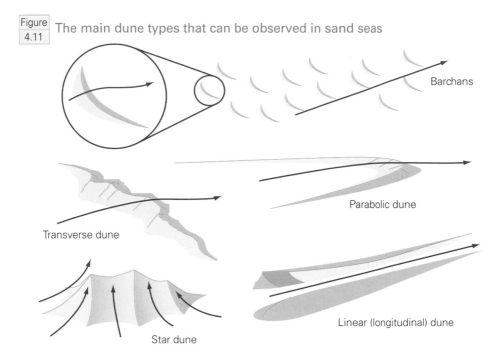

Figure 4.11 The main dune types that can be observed in sand seas

Barchans

Parabolic dune

Transverse dune

Linear (longitudinal) dune

Star dune

30° and 35° north and south of the Equator, where the prevailing winds blow from different directions in summer and winter.

Many different dune classifications, based on shape, orientation or number of slip faces, have been proposed. McKee (1979) developed a morphological classification that grouped dunes according to their shape and the number of slip faces into five main categories: crescentic, linear, reversing, star and parabolic.

Crescentic dunes

Crescentic dunes are the most common type. They are mounds of sand that are crescent-shaped in plan view, with a width greater than the length. The concave side has a slip face.

Barchans are crescent-shaped with two horns that face downwind. They are formed by a prevailing wind that comes from just one direction. The horns move faster than the main body of the dune, since the wind does not have to move as much sand (Figure 4.12). The windward slope is gentle, while the leeward slope is much steeper. Saltation builds up the windward side. On the leeward side the sand flows, slides and slumps down the steep slope. Barchans are highly mobile and can move across the desert surface quickly — at speeds of over 30 m a year.

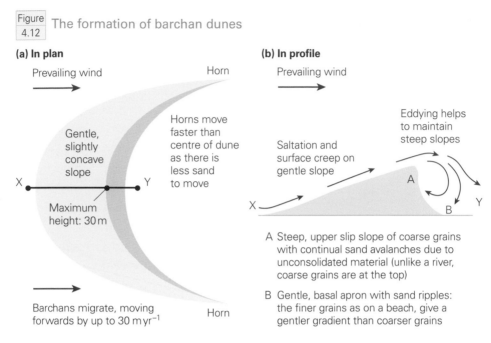

Figure 4.12 The formation of barchan dunes

(a) In plan

Prevailing wind

Horn

Gentle, slightly concave slope

Horns move faster than centre of dune as there is less sand to move

X ●————————●————— Y

Maximum height: 30 m

Barchans migrate, moving forwards by up to 30 m yr⁻¹

Horn

(b) In profile

Prevailing wind

Saltation and surface creep on gentle slope

Eddying helps to maintain steep slopes

A

X

B

Y

A Steep, upper slip slope of coarse grains with continual sand avalanches due to unconsolidated material (unlike a river, coarse grains are at the top)

B Gentle, basal apron with sand ripples: the finer grains as on a beach, give a gentler gradient than coarser grains

Transverse dunes are found in large erg environments and can be several hundred metres high. They develop at right angles to the prevailing winds. They have steep and gentle sides that form in a similar way to the barchan dunes.

Linear dunes

In plan form, linear (longitudinal) dunes are straighter than the crescentic dunes. They form parallel to the prevailing wind direction. They can be more than 100 km in length, and tend to be much greater in length than width. The slip faces are found on alternate sides. Linear dunes occur either as isolated ridges or in groups of parallel ridges, and cover a larger area of desert than any other type of dune.

There are two distinctive types of linear dune: linear ridges and seif dunes. **Linear ridges** are simple elongated dunes with crests that are influenced by the prevailing wind direction. **Seif** dunes have serrated crests and sides. Their shape is determined by localised wind eddies. They can be 100 km long and 200 m high.

Reversing dunes

These are ridges of sand that have a winding pattern. They occur in areas where there are two strong seasonal winds blowing from opposite directions — for example, northeasterlies may dominate in summer and northwesterlies during winter. Reversing dunes are essentially variants of other dune types, modified by the two different wind directions.

Star dunes

Star dunes, also known as **rhourds**, develop where strong winds blow from several different directions, but there is no one dominant wind direction. They are pyramidal in shape, and have slip faces on three or more 'arms' that join together in a higher central mound of sand. They grow upwards, and may form some of the tallest dunes on the planet. Where a sequence of star dunes merge together, they form a long and serrated ridge, known as a **draa**.

Parabolic dunes

Parabolic dunes have a pronounced U-shaped plan form, with a convex nose and elongated arms. At first glance they have a similar shape to crescentic dunes, except that the arms point upwind. The arms are fixed in place by vegetation, while the rest of the dune is able to move forward. These dunes often experience 'blow-outs', where the sand at the centre of the dune is blown away, leaving behind a shallow depression.

Climbing/falling dunes

Dunes can cross ranges of hills or other rocky barriers. They develop in areas that are mainly rocky, with undulating relief, but that also have small amounts of sand that can be moved by strong winds. Where the sand can be seen piled in a mound and being blown up a rocky barrier, it is known as a climbing dune. Where the sand is being transported down a rocky barrier onto a lower area of relief, it is known as a falling dune.

Dune form

McKee observed that dunes may occur in one of three forms: simple, compound or complex.

- **Simple dunes**. These are basic dunes. They have the minimum number of slip faces needed for them to be classified as this particular dune type. They represent a wind regime that has stayed at the same intensity and direction since the dune was formed.
- **Compound dunes**. These are large dunes on the top of which small dunes of a similar type are superimposed.
- **Complex dunes**. These are combinations of more than one type of dune: for instance, a crescentic dune with a star dune on its crest.

Activity 2

Study Table 4.1, which shows the average wind direction for each day during June 2009 at Windhoek in Namibia. Figure 4.13 shows the location of Windhoek.

| Table 4.1 | Average wind direction, Windhoek, in June 2009 |

Day	Average wind direction (degrees)	Day	Average wind direction (degrees)	Day	Average wind direction (degrees)
1	335	11	006	21	016
2	334	12	020	22	359
3	333	13	017	23	340
4	292	14	007	24	312
5	321	15	006	25	252
6	337	16	002	26	167
7	287	17	313	27	173
8	199	18	072	28	158
9	158	19	093	29	142
10	034	20	047	30	150

Source: www.namibiaweather.info

(a) Use the data from Table 4.1 to construct a wind rose diagram for Windhoek.

(b) Explain how the wind directions shown in Table 4.1 might influence the formation and morphology of sand dunes.

(c) How do the data in Table 4.1 help to explain the star dunes found in the eastern Namib Sand Sea?

(d) Use the website **www.orusovo.com/guidebook/content3.htm** to research the main types of dune found in the Namib Sand Sea. Investigate the formation and characteristics of the dunes.

| Figure 4.13 | Namibia: the location of Windhoek and the Namib Desert |

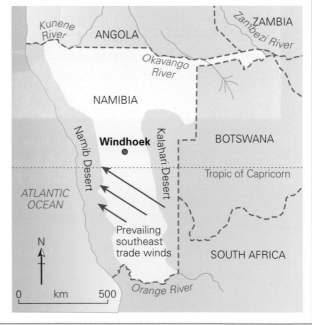

Water as a landforming agent

Whereas depositional features are more commonly seen in the lowland areas, water erosional features occur widely in desert mountains and plateaux.

Rainwater from low-intensity storms in arid and semi-arid environments will be lost to evapotranspiration or percolate downwards through the soil. However, when isolated heavy downpours occur, they lead to much greater overland flow and can have a considerable effect on the physical landscape. **Orographic uplift** (where air is forced to rise over mountainous areas) and convection currents can cause intense thunderstorms in desert areas.

Most streams and rivers found in these regions are ephemeral (only flowing during a period of heavy rain and for a short time afterwards). Although surface runoff over the course of the year is limited, when it does occur it has enormous power. Runoff after a heavy storm is rapid because of:
- high-intensity rainfall
- sparse vegetation cover that allows little interception of water
- high channel density (e.g. frequent rills and gullies of steep unvegetated slopes)
- hard dry sun-baked soil that limits infiltration
- rainsplash that fills soil pores quickly, causing soils to lose permeability
- shallow soils allowing little water storage

Fluvial erosion

Landforms created by water erosion include **wadis**, **canyons**, **pediments** and other features such as **mesas**, **buttes**, **spires** and **inselbergs**.

Wadis

Wadis are dry river beds that form temporary channels following periods of rain. They are common in arid and semi-arid environments. Most rivers and streams in desert areas are ephemeral: they flow during flash floods or periods of high seasonal rainfall, and are short-lived.

Wadis have steep sides and a wide channel floor (Figure 4.14) covered with fluvial sediments. They range in size from small channels of no more than a few metres long to networks hundreds of kilometres long. Larger wadis may be relict landforms from past climates when there was more rainfall. High discharge, combined with steep channel gradients, means that ephemeral rivers have a lot of energy for erosion and transportation. The high volumes of sediment transported scour the river channel. Storms in desert areas are normally fairly localised, and may not affect the entire length of a wadi channel. As the river in the active part of the wadi flows, some water from the channel will evaporate or infiltrate and so discharge normally declines downstream. As discharge decreases, the rivers lose competence and **braiding** occurs.

Wadis are the main landforms of desert areas that result from water erosion. Smaller channel incisions carved by water form tributaries to wadis — these landforms are called **gullies** and **rills**. They become active during rainstorm events, and experience **headward erosion**.

Figure 4.14 Wadi in the Atlas Mountains, Morocco

Lucy Cole

Canyons

Canyons are desert gorges. They are much deeper versions of wadis, with steep (often vertical) sides cut into resistant rocks. Heavy rainstorm events and resulting rapid runoff create powerful flash floods that give streams a lot of energy to erode the land. Rivers flowing in canyons, for example the Colorado River, are usually **perennial**. They cut gorges down into solid rock. Canyons are found in mountainous and plateau areas in both arid and semi-arid environments. Because of the arid conditions, weathering on the valley sides is limited, allowing them to retain their steep, nearly vertical, slopes.

Canyons are formed through a combination of water action (i.e. vertical fluvial erosion) and tectonic uplift. The largest canyons on the Colorado Plateau, such as the Grand Canyon, have formed in this way. Abrasion is the principal erosional process, with erosion being predominantly vertical rather than lateral because of the solid rock walls of the channel. The coarse rock debris scours out the channel vertically as it is transported. The river discharge within canyons will vary as a result of seasonal rainfall and flash flooding. During times of low discharge, channel braiding within the canyon will occur. Rapids form where swollen tributary streams carry boulders into the main channel.

Rock type also affects the slope profiles of canyons:
- **Slot canyons** have vertical sides. They are formed from a resistant rock type, which is similar throughout the cross-section.

- Stair-like canyons form where the cross section contains rocks of different resistance. Resistant rock forms areas of cliffs, whereas the weaker rock has a lower slope profile, as for example in the Grand Canyon (Figure 4.15).

Figure 4.15   The Grand Canyon

Lucy Cole

Activity 3

Examine the importance of tectonic processes in the evolution of deep canyons, such as the Grand Canyon, in the southwest USA.

The following websites are useful starting points:

geomaps.wr.usgs.gov/parks/province/coloplat.html

www.absoluteastronomy.com/topics/Geology_of_the_Grand_Canyon_area

www.bobspixels.com/kaibab.org/geology/gc_geol.htm

Pediments

Pediments are gently sloping erosional rock surfaces, with an angle of less than 7°, found at the base or foot of mountain ranges, cliffs or steep hills in the desert. They are one of the most typical landforms of arid and semi-arid environments. Sometimes pediments comprise bare rock surfaces, but for the most part they are covered with debris either from rockfalls or from alluvial fans.

Pediments are formed by the **parallel retreat** of the steep slopes of a plateau or mountain front (Figure 4.16). As these areas retreat, they leave behind a gently sloping rock platform. The slope angle remains almost constant, since the plateau retreats parallel to itself through weathering. Parallel retreat occurs because rock is removed from the base of steep slopes at the same rate that it is weathered. The fact that features such as mesas, buttes and spires (see below) all have similar slope profiles despite being at different stages of erosion, provides further evidence of parallel retreat. Weathering is most effective where harder rock is underlain by softer rock, allowing the harder rock to be undercut.

| Figure 4.16 | Parallel retreat and the formation of pediments |

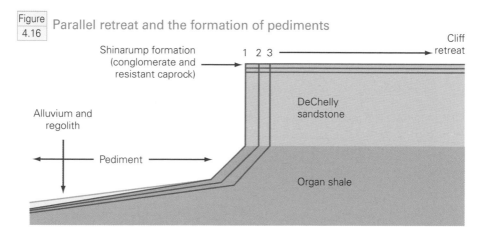

Weathering and sheet runoff work together to form pediments. The debris that builds up from weathering of the slopes of mesas and buttes is washed away by sheetflow, and the rock surface is scoured by the debris carried down the slope by water.

Mesas, buttes and spires

Mesas, buttes and spires are examples of relict hills that have been isolated by weathering and water erosion. The desert surface, incised by wadis and canyons over millions of years, is left with isolated plateaux and **mesas** with flat tops and steep sides. Rock pediments, covered with scree from weathering and rockfalls, often form at the base of mesas. As water and wind continue to erode the mesa, it becomes smaller in size until it is no longer wider than it is tall. At this point it becomes known as a **butte**. Erosion of the butte continues until a thin pillar-like formation known as a **spire** is left. Eventually the spire will be eroded away.

Mesas, buttes and spires are the remains of much larger plateau surfaces shaped over time by weathering, erosion and reduced by parallel retreat. They develop in sedimentary strata with horizontal bedding planes, capped by a more resistant rock. A classic landscape of mesas, buttes and spires can be seen in Monument Valley in Arizona and Utah (Figure 4.17).

Figure 4.17 | Mesas, buttes and spires in Monument Valley, southwest USA

Inselbergs

Inselbergs are isolated relict hills or domes of resistant rock that are more rounded than the mesas and buttes. They are commonly found in the semi-arid regions, and tend to form in crystalline rocks such as granite.

Inselbergs formed at a time when rainfall was much higher than today. Deep chemical weathering occurred in the rock, and surface deposits were subsequently removed by water. They may be formed by the parallel retreat of slopes, or where the basal surface is exposed owing to surface stripping. Inselbergs are left 'stranded' (like rocky islands) as the level of the surrounding landscape is lowered by weathering and erosion.

The rounded appearance of inselbergs is due to the removal of the surface layers of rock over time by erosion and weathering. This relieves the pressure ('unloading') on the lower layers of rock, allowing them to expand parallel to the surface. The process of **pressure release** creates lines of weakness (pseudo-bedding planes) that are vulnerable to other weathering processes.

Fluvial deposition

Seasonal temperature fluctuations affect the amount of water present in arid areas and, therefore, the landforms that develop. Fluvial deposition creates a variety of features, such as **alluvial fans**, **bajadas**, **salt lakes** and **salt pans**. These features develop on low-lying areas of flat ground, and are often found within basin areas — broad depressions in the landscape and the focus of internal drainage.

Alluvial fans

Arid environments are prone to flash flooding, facilitating the development of alluvial fans; they also have hill-slopes that erode easily, thus providing plenty of sediment for deposition. Where perennial or ephemeral streams with large sediment loads emerge from a mountainous area onto a gently sloping lowland plain or basin, the stream will slow down and spread out into a fan shape formed by small distributaries. As it loses velocity and energy, it quickly deposits its sediment load. Deposition is further aided as the stream is no longer confined by steep-sided valley walls. The sediment deposited in the area at the foot of the mountain front forms a fan or cone-shaped feature known as an alluvial fan. These can be just a few metres or several kilometres wide. There is often a clear grading of sediment on alluvial fans: the upper part of the fan has the coarsest sediment, which is deposited first; smaller material is carried further away towards the foot of the fan. Some of the best examples of alluvial fans are found in Death Valley in California.

Bajadas

Bajadas are created where a number of parallel wadis arrive at the mountain front, forming a series of alluvial fans that join together to create a continuous apron of sediment covering the lower slopes. These features can be found downslope of desert mountain ranges (Figure 4.18). Sediment sorting on bajadas is similar to that on alluvial fans.

Salt lakes and salt flats

Hollows or depressions on the land surface may be occupied in part by shallow, ephemeral saline lakes (Figure 4.19) or **playas**. Water in the lake builds up through a combination of precipitation, groundwater and surface runoff

Figure 4.18 Alluvial fans joined together to form a bajada, Death Valley, California

Lucy Cole

from higher ground following rainfall. Salt lakes may also form where deflation hollows have been eroded downwards until they reach the water table. Playas usually have no surface outlet, and water remains on the surface temporarily until it is evaporated in the high temperatures. During periods when playas dry up, the clay bed floor of the lake cracks in a polygonal pattern (Figure 4.20). Salt lakes vary in size. Large playas in Death Valley occupy several square kilometres.

As the water from the lake evaporates, it leaves behind deposits of soluble salts that were carried in solution by the desert streams. These salts accumulate on the surface. Salt flats, or **pans**, are flat expanses of ground encrusted with salt and other minerals, often tessellated into

Figure 4.19 Playa lake at Badwater Basin, Death Valley, California

Lucy Cole

Figure 4.20 Dry clay floor, Death Valley

Lucy Cole

polygons (Figure 4.21). The most common salt is sodium chloride, although gypsum, sodium sulphate, magnesium sulphate and other salts can also be found. The salts in these pans can be exploited for commercial use. There is very little vegetation cover, although the edges of the pan will be occupied by halophytic plants (adapted to tolerate salt) such as saltbush. Mesquite bushes can be found where the water table is near the surface. Salt pans are susceptible to wind erosion: salt weathering loosens material on the floor of the pan, and this material can then be eroded by the wind.

There are two levels to salt lakes. The upper level contains water only after extreme rainstorms. Most of the time it consists of bare salt flats. The lower part has water in it more often, normally in the centre, and so will have a thicker build-up of salts. In past, more humid climates, permanent lakes occupied these areas.

Salt lakes have a variety of names, as shown in Table 4.2.

Figure 4.21 Salt pan and salt polygons at Badwater Basin, Death Valley

Lucy Cole

Table 4.2 Salt lakes

Name for salt lake	Where used
Playa	North America
Salar	South America
Chott	North Africa
Sabkhah	Arabian Peninsula
Kavir	Iran
Takir	Central Asia
Pan	Australia, South Africa

Oases

An oasis is a small, fertile area that is surrounded by desert. Oases are formed where the water-bearing rocks are exposed at the surface — often as a result of deflation.

Badlands in semi-arid regions

Badlands are areas where softer sedimentary rocks crop out and have undergone extensive water erosion. They are associated with the more humid, semi-arid environments, where the landscape is shaped by water action, rather than with the hyper-arid areas. Badlands occupy only small areas within desert regions — 2% of the Sahara, for example. They also occur in scattered locations outside the main desert regions.

Badlands develop where the rocks are soft and largely impermeable. Although rainfall is erratic in the semi-arid regions, it is often heavy, and runoff is rapid. Bedrock materials such as clays are easily eroded, the drainage density is high, and the sparse vegetation cover is insufficient to hold the bedrock materials together. Ephemeral streams are powerful enough to create some dramatic landforms.

The term 'Badlands' arose in the USA, where it referred to such areas' unsuitability for agriculture. These areas are characterised by steep slopes, loose dry soil, clays and deep sand. They have intensive, infrequent rainstorms, sparse vegetation cover and soft sediments, all of which contribute to high rates of erosion. The best known example is in the Badlands National Park in South Dakota.

Common landforms found in badlands include pipes, canyons, ravines, gullies and hoodoos.

Activity 4

Figure 4.22   A deeply gullied landscape at Zabriskie Point, Death Valley, carved into soft sedimentary rocks

Lucy Cole

(a) Sketch the badlands landscape shown in Figure 4.22 and add labels to show the rills, gullies and wadis.

(b) Describe and explain the shape, pattern and density of channels in the landscape shown in Figure 4.22.

Pipes

Pipes are eroded passageways that occur when water infiltrates through cracks when clays dry out. The water passes underground and, as it does so, it erodes the pipes. In subsequent rainstorm events, the water then runs through these pipes and enlarges them. Micro-piping is only a few millimetres in diameter. It can be identified through a honeycomb appearance to surfaces. Meso-piping can be up to a few metres in diameter. Roof collapses to the tunnels form discontinuous gullies.

Caves and natural arches

Larger areas of piping can lead to the development of caves. Over time, the caves may be eroded backwards to form natural arches. After heavy rainstorms, large amounts of water may run through the natural arches.

Wadis and gullies

In badland areas, wadis are steep, with large amounts of debris occupying the wadi floors. There are often many tributary gullies flowing into them. Gullies are small valleys that have been eroded by running water after heavy rainfall. These experience headward erosion and cut into and undermine hill slopes. This can lead to slope failure and slumping in the hillsides.

Relict hills

Water action over time leads to greater dissection of the badlands, and the formation of relict hills. In younger badland areas, these hills may be hundreds of metres high. In lower-lying areas where they have experienced a longer period of erosion, they may be reduced to just a few metres high. The relict hills tend to be rounded, and incised by numerous rills.

Hoodoos

Where softer rocks are overlain by a more resistant cap rock, the relict hills may develop into pillar-like structures known as 'hoodoos'. Hoodoos are composed of soft sedimentary rock with a harder cap rock that protects the column of rock below. The pillars have a variable thickness, as the alternating bands of rock are eroded at different rates. The most famous area of hoodoos is in the northern part of Bryce Canyon National Park in Utah (Figure 4.23). The rocks in Bryce Canyon include limestone, siltstones and mudstones, capped with a magnesium-rich limestone called dolomite. The dolomite is more resistant to chemical weathering, and protects the weaker limestone underneath.

The hoodoos in Bryce Canyon are eroding at a rate of between 0.6 m and 1.2 m every 100 years.

Figure 4.23 Hoodoos at Bryce Canyon

Michael Raw

Activity 5

Use the website **www.brycecanyon.com** to help you complete the following tasks.

(a) Draw an annotated diagram to show the hoodoos at Bryce.

(b) Label the different rock types, and indicate how these have influenced the shape of the landforms.

(c) Describe the sub-aerial processes that influence the formation of hoodoos.

5 Arid ecosystems

Ecosystems survive in hot arid and semi-arid regions despite the very harsh conditions for plants and animals. Climatic extremes have led to the development of finely balanced ecosystems adjusted to:

- high summer temperatures
- low and unreliable rainfall
- unstable surfaces
- strong winds
- thin, infertile and saline soils

Water is crucial to the survival of all living things, and the desert species have managed to adapt in remarkable ways. Seasonal change involving long-term drought and short-term abundance of water is the major force driving evolution, natural selection and adaptation in these regions.

Plant adaptations

While deserts are arguably the harshest of any natural environment, at least 75% of the Earth's desert area supports some form of vegetation. The only areas where vegetation is absent are the ergs (large sand seas), which are constantly on the move (e.g. the Chech Erg in the Sahara), and the extreme arid environments of the Atacama and Namib deserts.

Variations in temperature, rainfall, soils, rock type, altitude and relief can lead to wide differences in biodiversity both within and between arid areas. For example, the Sonoran Desert has a diverse range of vegetation, while in the Atacama Desert biodiversity is much smaller. Despite the vast contrasts in vegetation between the different parts of the world, however, some generalisations can be made. First, where there is vegetation, **plant succession** does not develop towards a 'climatic climax' state, because of unreliable rainfall and constantly changing environmental conditions. Second, many desert plants are small. Third, the vegetation cover occurs in patches rather than as continuous cover, with 'stands' of vegetation consisting of just one species (Figure 5.1). Fourth, groups of species grow close together, relying on each other for shade. Lastly, the vegetation does not form structured layers (as in more humid environments) unless a permanent water source is available.

Figure
5.1 Stands of creosote bushes at Stovepipe, Death Valley, California

Plants have adapted to the arid and semi-arid conditions by developing:

- deep or wide root systems to maximise access to moisture
- short life cycles (following the sporadic rainstorms)
- the ability to store moisture in their stems or leaves
- physical characteristics to prevent water loss

Plants in arid areas can be placed into three main groups according to how they adapt to aridity and drought: **xerophytes**, **ephemerals** and **halophytes**.

Xerophytes

Xerophytes are plants adapted to withstand drought. Adaptations include succulence (storage of water in their tissues), reduced transpiration and long root systems.

Succulents

Succulents have modified physical structures in order to resist drought. They store water in their stems, leaves and/or roots.

Cacti, indigenous to North and South America, are the best-known family of succulent plants. They cope with drought by storing water within their fleshy leaves or stems. Water absorbed during the seasonal rains is stored in specialised tissues of plant cells called **vacuoles**. The plant draws on this water during dry periods.

The saguaro cactus (Figure 5.2) found in the deserts of the southwest USA is one of the most impressive species of cactus. Plants grow up to 15 m tall and can live for 200 years. After rainfall, a large saguaro cactus can absorb up to 8000 litres of water which it stores in its fleshy stem and arms.

Cactus plants have extensive shallow roots radiating in all directions, allowing them to absorb large amounts of water in a short time when it rains. They can also store water within bulbs on their root systems, and can survive years of drought by relying on the water they receive during a single rainfall event.

In Africa, Asia and Australia, species of euphorbia (Figure 5.3) and aloe are the main types of succulent.

Reducing transpiration

Many plants survive drought by keeping transpiration levels to a minimum. They have thorns or spikes instead of leaves, or only very small leaves, to reduce loss of moisture through transpiration. Grasses have narrow leaves that can roll inwards. Some plants lose their leaves completely in the dry season but have green stems that allow photosynthesis to continue.

Figure 5.2 — Saguaro cactus growing in the Mojave Desert, southwest USA

Lucy Cole

Figure 5.3 — Euphorbia

Corel

Other adaptations to minimise transpiration include stomata which close during the day and open at night when it is cooler, as in cacti. Stomata are pores in the leaves and sometimes the stems of plants. They absorb carbon dioxide used for photosynthesis, and release water that is lost to evaporation (i.e. transpiration). By keeping their stomata closed during the heat of the day, plants are able to reduce their transpiration levels to a minimum. Stomata also occur in grooves on leaf and stem surfaces, restricting the loss of water vapour by creating a moist microclimate within the grooves. Many succulents have crassulacean acid metabolism (CAM), which allows them to carry out photosynthesis even during the day when their stomata are closed. Carbon dioxide is taken in at night and then stored until daytime when photosynthesis can occur. Stomata are often more numerous on the underside of the leaf, since less evaporation will take place in the cooler, shady conditions.

Many desert plants have thickened, waxy leaves or cuticles. The non-porous wax acts like a waterproof layer to reduce the loss of moisture and at the same time helps to reflect some of the sun's heat.

Phreatophytes

Phreatophytes are a group of plants with very long roots which allow them to access water from deep below the surface. They are most commonly found in areas where the water table is closer to the surface (e.g. in the vicinity of wadis). The roots of some phreatophytes, such as mesquite, can tap water 15–20 m underground. Mesquite are native to the southwest USA, where they are found in abundance. Even more remarkably, the roots of the zizyphus lotus in southern Morocco reach depths of 60 m. Because these plants are able to utilise water sources found deep underground, they continue to photosynthesise throughout the dry summer months.

The creosote bush (Figure 5.4) has adapted successfully to life in desert areas. It has a root system that reaches deep groundwater, as well as a second, radial root system. Its other adaptations include a smell and a taste that deter animals from eating it (hence the name) and small leaves with stomata that close during the day to avoid moisture loss and open at night to absorb moisture.

The longer roots of phreatophytes enable them to survive in constantly shifting sand dune areas. Their long, tough roots hold onto the dunes and anchor the plants down.

Figure 5.4 Creosote bush in Joshua Tree National Park, southwest USA

Lucy Cole

Ephemerals

Ephemerals are plants more commonly found in arid environments where the rainfall has some seasonal pattern. They can be **annuals** that have a brief life cycle or **perennials** that have a brief period of activity.

Annual ephemerals escape drought by remaining as seeds until rain falls. They are small plants with shallow root systems and can survive for many months or even years in a desiccated state, only loosely rooted to the desert surface. Typical annual desert plants germinate after heavy seasonal rain, flowering, seeding and dying within a few weeks or even days, depending on the weather and climatic conditions. Most annuals bloom during the desert spring, when they create a brief explosion of colour. The resulting seeds will then lie dormant until the next rain.

Perennial ephemerals do not die after flowering and seeding; instead, to survive extremes of heat and lack of moisture, they become dormant during periods of drought or extreme temperature. They grow only when conditions are at an optimum.

The desert lily, native to the Mojave and Sonoran deserts of California, Arizona and Mexico, is a perennial that stores energy in its roots and bulbs, which are formed deep in the ground. The energy that it stores in spring enables it to survive most of the year in a dormant state. In the summer, the above-ground parts of the plant dry out so that none of the plant is exposed to the heat and drought. The plant will then spring to life again when water becomes available.

The desert paintbrush (Figure 5.5), native to the deserts and mountains of the western USA, is a perennial shrub that germinates in the spring following winter rains. It grows quickly and then flowers and produces seeds. The seeds are hardy and are able to resist drought and very high temperatures.

| Figure 5.5 | Desert paintbrush |

Lucy Cole

Even the most barren-looking desert soil contains dozens of seeds of annual and perennial plants waiting for the winter rains. Soil in the Sonoran Desert has recorded seed densities of up to 10 000 seeds per square metre. Seeds can survive without water indefinitely and only begin to germinate when the rain washes away anti-sprouting chemicals on their shells.

Halophytes

Halophytes are plants that have adapted to conditions high in salt.

Saline habitats occur in lowland areas of arid and semi-arid environments, due to low levels of precipitation combined with high rates of evaporation. These conditions can create salt pans and so the plants growing there must be tolerant of saline soils. Salt is toxic to most plants and especially so when the soil water concentration is relatively low. Because saline areas are extreme environments, species diversity is low — sometimes restricted to a single dominant species (e.g. saltbush). Plants adapted to saline soils and low rainfall are some of the hardiest of all desert species.

Halophytes use many different strategies to survive:

- The saltbush excretes salt from salt glands onto its leaves when conditions become too saline.
- Succulents such as pickleweed have a high water uptake, which compensates for the high salt content.
- Some halophytes' seeds can germinate in the presence of high salinity.
- Seeds of other halophytes are kept dormant due to the high salinity, but germinate when the rains come and reduce the salinity on the seeds.
- Some root systems of halophytes can exclude salts.
- The salt is compartmentalised in certain tissues of some halophytes, which act as salt storage away from growing cells.
- Some halophytes develop succulence, which dilutes the level of salt in the plant and stores water for use during dry periods.

Other plant adaptations

Food for animals is scarce in arid areas owing to the harsh climate. Many plants have adapted to protect themselves from being eaten. The prickly pear cactus (Figure 5.6) and the acacia tree have spines and thorns. The sodium apple contains a bitter and poisonous latex.

Because competition for water is intense in arid environments, plants prevent other plants from

Figure 5.6 Prickly pear cactus

Lucy Cole

intruding on their growing space. The creosote bush gives off toxic chemicals that prevent other plants from growing nearby and competing for the same local water supply. This system of survival, known as **allelopathy**, is responsible for the uniform spacing of plants in desert areas.

Some plants are **parasitic** and take water and nutrients directly from others. For example, the orobanca, found in north Africa, lives off small succulent bushes growing in sand dunes.

Seed dispersal is difficult when suitable sites for plant growth occur sporadically. Three mechanisms are available: wind dispersal, dispersal by animals and dispersal by water. Some plants form seeds with wings or parachutes or cottony masses so that they can be carried a long way by the wind. Tumbleweed is adapted to break off at ground level and is then blown along the ground by strong gusts of wind, scattering its seeds as it travels. Cholla, or jumping, cacti (Figure 5.7) have barbs that can attach themselves to the coats of passing animals. Even in the desert, some seeds are water-dispersed by flash floods. The blue palo verde tree grows in washes in the California desert; it has hard seeds that need to be scarified or abraded as they are carried along in a flash flood before they germinate so that they do not open before they have been dispersed.

Figure 5.7 Cholla cacti, Joshua Tree National Park

Lucy Cole

Activity 1

Use the following websites to find examples of plants that survive in the Sonoran Desert.

www.blueplanetbiomes.org/sonoran_desert.htm

www.desertusa.com/flora.html

Select three plants, and annotate a photo or sketch of each of these plants to show how it has adapted to the arid conditions.

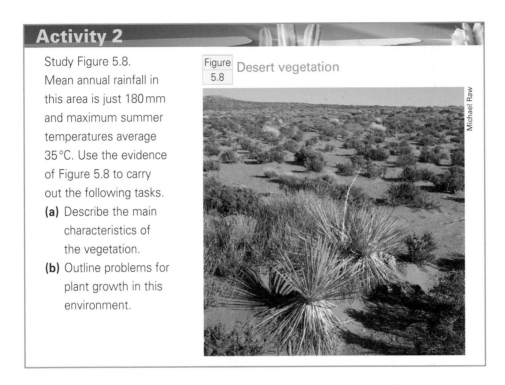

Activity 2

Study Figure 5.8. Mean annual rainfall in this area is just 180 mm and maximum summer temperatures average 35°C. Use the evidence of Figure 5.8 to carry out the following tasks.

(a) Describe the main characteristics of the vegetation.

(b) Outline problems for plant growth in this environment.

Figure 5.8 Desert vegetation

Michael Raw

Animal adaptations

The desert is an even more inhospitable environment for animals than it is for plants. Animals are heated directly by radiation from the sun, and indirectly by conduction from rocks and soil and by convection from the air. They can only function within a narrow temperature range (the range of **thermoneutrality**). In addition to extremes of temperature, animals face two other environmental challenges in hot arid and semi-arid environments: lack of water and shortages of food.

Despite these problems, many species of animal have successfully adapted to survive in hot arid and semi-arid environments. In terms of survival, animals have one big advantage over plants — they can move. Animal strategies for survival take two forms: behavioural modifications and physiological adaptations. Each has evolved over many generations.

Behavioural modifications

Animals may modify their behaviour in response to diurnal changes in temperature or seasonal climatic changes.

Migration

Many birds and most large mammals living in arid areas avoid the most extreme seasonal temperatures and drought. They migrate across desert plains or into the mountains.

Herbivores such as the red kangaroo of Australia and desert antelopes like the addax of north Africa are desert animals adapted to cover large distances very quickly. This enables them to move between areas of vegetation and follow the rains. Because mobility is crucial, the kangaroo carries its young in its pouch while an addax calf can walk within an hour of its birth.

Aestivation

Hibernation is a strategy used by animals to escape cold weather and food shortages during winter. In arid areas, some animals enter an equivalent state called **aestivation** during the extreme hot and dry conditions of summer. During aestivation, many physiological changes take place, including lowered heart rate (by as much as 95%) and lowered body temperature.

No desert mammal ever aestivates fully, because of the risk of desiccation or drying out. Animals that aestivate include reptiles like the North American desert tortoise, amphibians such as frogs, salamanders and spadefoot and cane toads, and other creatures like lungfish and spiders. Desert tortoises (Figure 5.9) in the Mojave Desert are most active from February to May, and spend the remainder of the hot, dry summer aestivating, normally in burrows. In the relatively cooler Sonoran Desert, the monsoon rains allow desert tortoises to be active for longer — from July to October.

| Figure | The desert tortoise aestivates |
| 5.9 | over the hot, dry summer months |

Dan Suzio/SPL

Nocturnal activity

Small animals cannot migrate long distances but they can avoid the hottest part of the day by being active only at night when it is cooler. Owls and bats sleep during the day and hunt at night. The fennec fox is a small nocturnal fox of the

Sahara Desert. Its ears are so sensitive that it can hear its insect prey walking on sand at night. It also has thick fur on the soles of its feet to protect it from hot sand, large ears that also serve to dissipate heat, and a pale, reflective coat.

Avoiding midday heat

Almost all desert animals stay out of the sun during the hottest part of the day. Some are not strictly nocturnal and remain awake during the day, but become active only during the cool of the morning or at dusk. For the remainder of the day they rest in cool, moist burrows or seek the shade of rocks.

Lizards enjoy heat and will bask in the sun during the day and burrow into the sand for warmth at night. When the sand becomes too hot for their feet, many lizards execute the 'thermal dance'. The lizard props itself up on its tail and lifts one front foot and one back foot, holding them up in the cool air. It then alternates them with the opposite pair. By avoiding full contact with the hot ground surface, the lizard can cool its feet by as much as 10 °C in a few seconds. If the heat becomes more extreme, the lizard retreats below the surface to seek a cooler environment.

Some snakes move sideways rather than forwards, thereby minimising the area of skin in contact with the hot sand. Sidewinding keeps most of the body above the surface and allows the snake to move without overheating. An example is Peringuey's adder, also known as the sidewinding adder or dwarf puff adder, a venomous viper of the Namib Desert.

Solifugae

Solifugae are large carnivorous arachnoids that are themselves a source of food for many animals. They inhabit most of the world's deserts, though are absent from Australia. 'Solifugae' means 'those that flee from the sun' in Latin. Vernacular names include sun spider, camel spider, wind scorpion and sun scorpion. These creatures have adapted to be unusually tolerant of heat and drought. During the hottest part of the day, they either retreat to a cool burrow or avoid the surface heat by jumping up into vegetation to cool off momentarily. Temperatures just above the ground can be as much as 15 °C cooler than on the surface.

Physiological adaptations

Physiological adaptations are evolved structural features that result in an improved ability of the animal to cope with the environmental conditions and enhance its survival.

Conserving water

Animals that live in arid areas have evolved many ways to reduce their water requirements, to conserve water in their bodies or to obtain fluid from unconventional sources.

- Most animals get their water from the food they eat — for example, succulent plants, seeds and the blood of their prey. Animals like the desert tortoise get nearly all the water they need from their food. They also dig shallow pits to catch water and will wait by these when rain is due.
- Many animals store water in their tissues.
- The outer skeletons of many insects and spiders are almost impervious to water, allowing them to minimise moisture loss.
- The hyper-arid Atacama Desert receives moisture through coastal fogs. Many reptiles and insects have physiological adaptations allowing them to intercept this fog water.
- Birds and reptiles excrete waste as uric acid, thereby losing very little water. Other animals may concentrate their urine in order to reduce water loss. The kangaroo rat has highly efficient kidneys that concentrate urine and produce dry droppings.
- Kangaroo rats reduce water loss from breathing by having specially adapted nasal cavities.
- Some animals take advantage of rainfall. Butterflies and bees time their emergence from pupae to coincide with the sudden crop of flowers that bloom after rain. Their young mature and develop to the pupa stage before the onset of drought.

Figure 5.10 Jack rabbits have evolved large ears to dissipate heat

Dissipating/reflecting heat

Animals of arid regions often have pale skin, fur, scales or feathers to reduce heat absorption. Some lizards, such as the chameleon, change their skin colour in order to reflect more heat during the day.

The desert fox and desert hare found in the Sahara have adapted to excessive heat by evolving enormous ears with blood vessels close to the surface. This gives the ears a greater cooling surface. Jack rabbits in North America (Figure 5.10) have similarly evolved large ears.

Wolfgang Kaehler/Alamy

Of the larger mammals, the camel is the best adapted to life in the desert:

- It can store large quantities of food within its hump as fat, which can then be metabolised to provide energy, enabling the camel to survive for several months without eating.
- It has a large capacity to take in water. It can drink up to 46 litres at a single session.
- It has thickened skin on the lips and in the mouth, allowing it to eat thorny xerophytic plants without feeling pain.
- It has long eyelashes to protect its eyes from sandstorms.
- It has well-padded feet to allow it to travel long distances over shifting sand dunes without sinking into the sand.
- It has thick fur and underwool to provide warmth by night and insulation against heat by day.
- It can close its nostrils to keep out sand.
- It has thick, leathery patches on its knees to prevent burns when kneeling on hot sand.
- It has long, strong legs, enabling it to travel long distances while keeping its body away from hot sand.
- It has hair inside and outside its ears to keep out sand.
- It has thick eyebrows to shield its eyes from the sun.
- Unlike most mammals, a camel's body temperature fluctuates throughout the day from 34 °C to 41 °C, enabling it to conserve water by not sweating as the environmental temperature rises.

Activity 3

Use the following website to help you research animals that live in the Sonoran Desert: **www.blueplanetbiomes.org/sonoran_desert_animals_page.htm**
Summarise the ways in which three different animals have adapted to drought.

Activity 4

Explain how human activities could affect the distribution of plants and animals in arid environments.

Soils

The soils in arid regions are generally infertile. The main exceptions are the soils found alongside permanent watercourses such as the Nile, or in oases where long-term irrigation and cultivation of the land has aided **pedogenesis** (the process of soil formation).

Three major factors restrict soil development in arid areas:
- lack of moisture (low and unreliable rainfall)
- high levels of evapotranspiration
- sparsity of vegetation cover

Aridisols

The **zonal soils** of hot arid regions are known as **aridisols** (Figure 5.11). Aridisols cover approximately one-third of the Earth's land surface. An aridisol is defined as a soil of dry regions that is alkaline or saline and contains only small amounts of organic material.

Figure 5.11 The profile of an aridisol

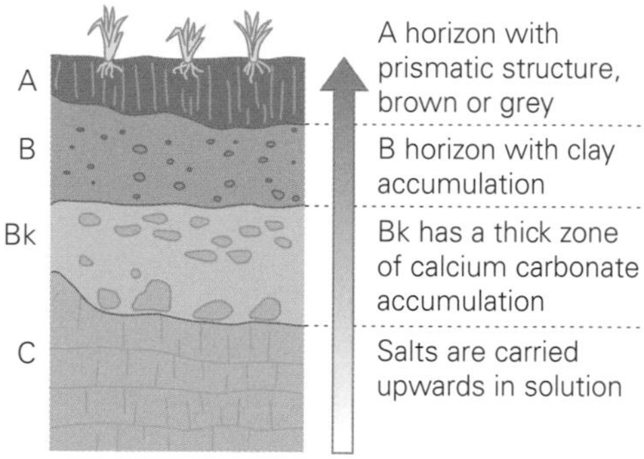

A horizon with prismatic structure, brown or grey

B horizon with clay accumulation

Bk has a thick zone of calcium carbonate accumulation

Salts are carried upwards in solution

Aridisols are slow-forming and occur mainly on gentle slopes. Because of sparse vegetation cover and low plant biomass, aridisols have a low organic content — typically just 1–2%. Even in areas of higher biomass, lack of moisture slows biological activity and rates of organic decomposition. Shortages of humus mean that soils are infertile and contain few of the nutrients needed for plant growth.

Aridisols range in colour from a yellowy red to grey-brown, and have a thin topsoil. The lack of organic content means that aridisols are mainly mineral soils. Arid zone soils tend to be shallow and coarse-textured, and develop slowly. Low rainfall limits chemical weathering of the bedrock, while leaching is rare, since rates of evaporation and transpiration exceed rainfall.

As a result, mineral ions accumulate in the soil. This causes high salinity and alkalinity, with pH values typically between 7.0 and 8.5 — levels that are lethal for most plants. The soils also develop hard, salty crusts that impede root development.

Arid zone soils are extremely fragile. The lack of vegetation cover exposes them to erosion by wind and water. Pedogenesis is slow and depends on long-term ground stability and minimum erosion.

In areas where mineral salts are more abundant, aridisols are replaced by **halomorphic soils** — either **solonchaks** or **solonetzes**.

Solonchaks

Solonchaks are soils with a saline horizon. They are found in areas with very high rates of evaporation, such as within mountain basins and on salt flats. They form on unconsolidated material — for example **loess** or **alluvium**. A high water table brings salt to the surface. Carbonates and chlorides form a thick crust, which makes the soil infertile. Soil horizons are hard to distinguish, and salinity decreases with depth.

Solonetzes

Solonetzes (also called burnout or gumbo soils) are soils that form in areas where there is sufficient rain to cause some leaching. However, the rain is not enough to wash the minerals right out of the soil. The surface layers are not saline but are shallow and their sparse organic content provides few nutrients for plants during the growing season. The subsoil (B horizon) is rich in sodium salts, which makes the soil hard to cultivate. The subsoil may form a tough impenetrable hardpan some 5–30 cm below the surface that prevents water from infiltrating.

Activity 5

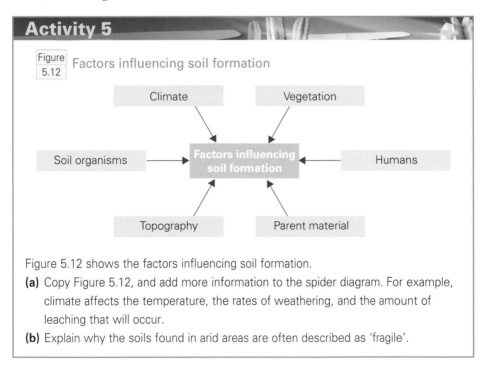

Figure 5.12 Factors influencing soil formation

Figure 5.12 shows the factors influencing soil formation.

(a) Copy Figure 5.12, and add more information to the spider diagram. For example, climate affects the temperature, the rates of weathering, and the amount of leaching that will occur.

(b) Explain why the soils found in arid areas are often described as 'fragile'.

Desert food webs

Desert food webs are often complex, owing to the seasonal nature of the vegetation. A lack of readily available food results in sparse animal populations. Desert areas have the lowest organic productivity of any **biome**: the average net primary productivity is just $90\,g\,m^{-2}\,yr^{-1}$. With low biomass, most food chains tend to be short, and rarely include more than three **trophic** (nutrition) **levels**. Few of the animals found in arid areas are specialist carnivores (Figure 5.13). The lack of reliable food sources means that many animals found at the higher trophic levels are omnivorous, taking advantage of any available food source.

| Figure 5.13 | A desert food web for the Mojave Desert, southwest USA |

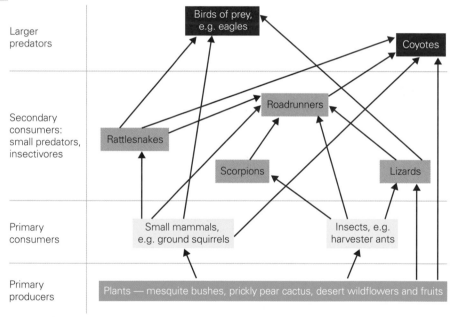

Fragile ecosystems

Arid zone ecosystems are finely balanced and easily damaged. Because of their low rates of net primary production, limited biodiversity and slow rates of recovery from damage, arid zone ecosystems are vulnerable to human activities. Even a slight disruption can cause irreversible change. Once the vegetation cover is lost, soil that has taken thousands of years to form is quickly eroded by wind and flash floods. Moreover, the harsh climatic conditions mean that nutrient stores and flows within the nutrient cycle are small and easily broken.

Human activities have adversely impacted the natural ecosystems of arid and semi-arid environments in a number of ways:

- Clearing vegetation for settlements, roads, reservoirs, mines etc. Natural ecosystems can be completely destroyed as a result.
- Collecting species of plants that have an economic value. This can upset the original balance of species. For example, 'cactus rustling' in the Chihuahuan Desert, encouraged by the booming trade in wild cactus for garden ornaments and desert landscaping, threatens many species.
- Replacing natural ecosystems with cultivation. The replacement vegetation is often at a higher density than the original plants.
- Pastoral farming, particularly in periods of drought. This leads to overgrazing and erosion.
- Overcultivation. This destroys the soil's structure and leads to erosion.
- Overuse of irrigation systems. This leads to **salinisation** (the build-up of salts in the soil) and land degradation.
- Over-extraction of groundwater for agricultural or domestic use. This causes a lowering of the water table.
- Introduction of non-native species that may compete with the original flora and fauna. For example, tamarisk was introduced to the North American deserts between 1899 and 1915 for erosion control, windbreaks and ornamental use. However, it is an aggressive invader, and has crowded out native plant species. Where large thickets of tamarisk have colonised the desert, it has lowered the water table, since it has long roots and consumes vast quantities of water.

Physical factors also make arid ecosystems vulnerable to change. For example, dust storms can bury vegetation; intense flash flooding breaks the stems and foliage of vegetation, and removes plants from the sides of the swollen rivers and wadis.

Climate change poses another threat to plants and animals. Although indigenous plants and animals have evolved mechanisms to cope with heat and dryness, they suffer when they are exposed to long periods of drought. Regeneration is difficult because of the harsh conditions.

Global warming is likely to have an adverse impact on many arid ecosystems in the future.

Activity 6

Explain why desert ecosystems are fragile and so easily degraded by human activities.

6 Desertification

Definition and overview

Whereas aridity defines the state an environment is in, desertification is the process by which arid areas grow. The UNCCD (United Nations Convention to Combat Desertification) defines desertification as 'land degradation in arid, semi-arid and dry sub-humid areas resulting from various factors, including climatic variations and human activities'. Desertification refers to the expansion of desert-like conditions to parts of the world where they should not occur according to the climatic belt; desertification occurs mainly in semi-arid lands that border desert areas. There is a loss of the biological potential of the land, and desert-like landscapes and processes develop in semi-arid ecosystems.

Desertification is one of the most serious environmental problems of today. It is a global problem that could eventually affect more than a third of the Earth's land surface (more than 4 billion hectares) and more than a fifth of the world's population. Yet it is not a new problem. Ancient societies lost productivity in agricultural areas, most notably when shifting from hunter-gatherer economies to sedentary agriculture. Land degradation led to farm abandonment, population displacement and deserted villages.

The term 'desertification' originated in the 1930s following research including Stebbing's (1938) paper entitled 'The encroaching Sahara: the threat to the west African colonies'. This caused concern that sub-Saharan Africa could be buried by sand as the Sahara Desert moved southwards, and led to a call for international action. Later research revealed a lack of substantial evidence to back up the idea of 'advancing deserts'. Instead, there is evidence that human activity such as deforestation, overcultivation and overgrazing degrades land.

Two things in particular have brought awareness of contemporary desertification to the general public. After the First World War, farmers ploughed larger areas of the Great Plains in the USA because cereals could be sold for high prices following shortages during the war. Then, during the prolonged drought in the mid-west of the USA in the 1930s, the wind eroded large quantities of soil, causing a 'dust bowl' and a mass exodus by farmers. Second, the severe drought and famine in the Sahel from 1968 to 1973, which caused large numbers of people and animals to die of dehydration and starvation, was extensively reported by the world's media.

Causes of desertification

Land degradation in arid areas takes place in three main ways: soil erosion, salinisation and deforestation. The ultimate cause of desertification is the removal of vegetation from the land surface. Where the vegetation is removed, dry and unprotected soils are either blown or washed away, decreasing soil fertility. A number of factors are responsible for the loss of vegetation (Figure 6.1). The main ones are drought, climatic shifts, population pressure, poverty, and policy shortcomings.

Figure 6.1 The causes of vegetation removal

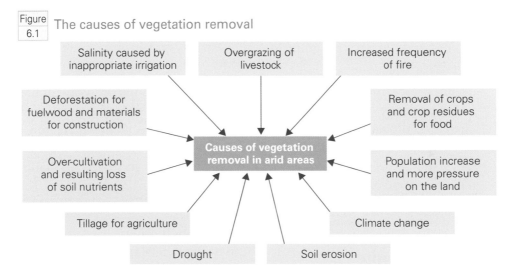

Drought

Drought is a natural phenomenon in arid environments, and occurs when the weather of an area is drier than normal. There are different types of drought, as shown in Table 6.1.

Drought is a short-term condition from which the natural environment can recover. Plants and animals respond quickly to variations in moisture availability. According to the UNCCD, the vegetation boundary to the south of the Sahara can move by up to 200 km depending on whether the year is wet or dry.

Droughts alone do not cause desertification, although there is a connection. Below-average rainfall has a direct impact on agriculture. As vegetative growth

Table 6.1 Types of drought

Drought	Description	Cause
Meteorological drought	A lack of rainfall, where precipitation falls below average for a prolonged period. The duration is no more than 10 years.	Short-term changes in the climate. These are normal occurrences, and the ecosystem will recover after the period of rainfall deficit.
Hydrological drought	A decrease in water levels in the rivers and lakes in arid areas, which may be caused by an increase in infiltration and/or evaporation	
Edaphic drought	A decrease in the infiltration capacity of the soil, meaning there is less soil moisture for seed germination and plant growth	Factors other than the climate
Agricultural drought	The needs of agriculture outstrip the availability of water. Alternatively, moisture may be available at an inappropriate time for crop growth.	

declines, areas may experience overgrazing by farm animals. With diminishing resources, farming may be forced to extend into more marginal areas. Therefore, drought and desertification are interlinked as land degradation is accelerated dramatically through mismanagement of the land. Well-managed lands can recover from drought, whereas land abuse and overexploitation during a period of drought increases the risk of permanent desertification. This process is difficult to reverse, and over time leads to a process of **aridification**, called desiccation, lasting for decades.

Climate change

Throughout history, deserts have grown and shrunk according to wetter and drier periods with no link to human activity. The scarcity of water in arid and semi-arid areas means that ecosystems that develop there are inherently unstable. Climate change is, therefore, a major threat to many semi-arid areas. Moreover, there is increasing evidence that climate change, due to global warming, will accelerate in the future.

Global warming will exacerbate the worldwide problem of desertification, and increase the total amount of land potentially at risk. Computer models make the following predictions:

- Many already dry areas will become even drier, particularly the semi-arid areas of Africa and south Asia. This could lead to a spread of desert.

- Global warming will increase the area of desert climates by 17% over the next 100 years.
- Rainfall in Turkey, Lebanon, northern Syria, Iran and Afghanistan will decrease by 30% by 2100.
- Desertification will enhance regional warming, through a variety of feedback mechanisms.
- Many arid areas will face low and increasingly erratic rainfall and a reduction in water resources, such as rivers and groundwater, as a result of increased regional temperatures.
- The variability of weather conditions and extreme events is likely to increase.
- Higher rates of evapotranspiration in the tropics and subtropics, combined with less reliable rainfall, may lead to an increase in both the frequency and intensity of droughts.
- Vegetation cover will decrease, and the type of vegetation may change, leading to a loss of biodiversity.
- Erosion of the land surface will increase as soils dry out and there is less plant cover to give protection from wind and water erosion.

Figures 6.2a and 6.2b show the forecast changes to global rainfall and global temperatures that would occur with a doubling of carbon dioxide in the atmosphere. However, the overall effects of climate change are not known with certainty. It is possible that some of the world's hot deserts may become wetter. This would be beneficial for the water supply and farming. The UNEP in 2003 noted that there is some evidence of re-greening of the Sahel.

| Figure 6.2a | Forecast changes in global rainfall in June–August with a doubling of atmospheric carbon dioxide |

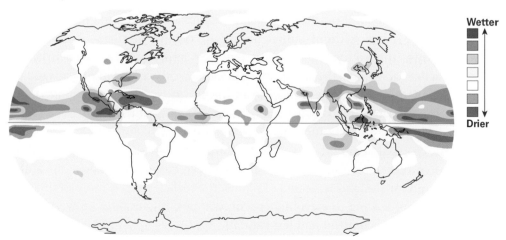

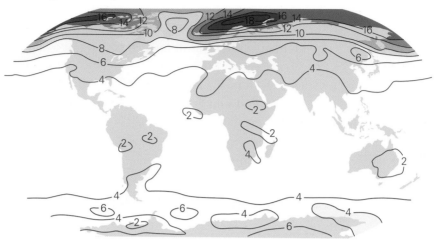

Figure 6.2b Forecast increases in global temperature in December–February with a doubling of atmospheric carbon dioxide

Increasing population pressure

Traditionally, people living in arid areas protected the biological and economic resources by following farming practices such as shifting cultivation and nomadic herding. The land has only low productivity and carrying capacity. Thus a sustainable agro-ecosystem can support only extensive cultivation, with low stocking densities.

However, in recent decades, population growth through both natural increase and migration has led to more settled communities, and a change in farming methods. During the 1990s, the arid and semi-arid areas experienced a growth of 18.5%, most of which was in LEDC countries. Growing human populations, and new agricultural practices, increase the pressure on natural resources:

- Farming crops without sufficient use of organic fertiliser, lack of crop rotation, the overuse of chemical fertilisers and badly managed irrigation practices all deplete the soil of nutrients. Poor soil conservation leads to land degradation. When the topsoil loses humus, it is more easily eroded, and new plants are less likely to establish. Incorrect irrigation practices can cause salinisation, which in turn can prevent plant growth. Soil quality, vegetation and crops are easily damaged during a period of drought. Removal of the native vegetation and fallow periods between crops both expose the soil to erosion. As the land is cleared of plants, the surface reflects more of the sun's radiation, leading to even higher temperatures, higher rates of evaporation, reduced condensation and less rainfall.

- Grazing animals removes native grasses. Traditionally, pastoralists moved animals in response to the patchy rainfall, and this prevented overgrazing. As populations increase, there are too many animals on too small an area of land, leading to overgrazing and the destruction of vegetation (Figure 6.3). As the number of edible perennial plants decreases, the proportion of annuals increases. The use of fences prevents domestic and wild animals from moving in response to the availability of food. Livestock compact the soil as they walk on it, which reduces the percolation rate of the soil, increases surface runoff and decreases the amount of water in the soil available to plants.

Figure 6.3   Overgrazing in the Masai Mara, Kenya

Michael Raw

- Firewood is the principal source of fuel in many LEDCs. Forests have been cleared in many semi-arid areas, exposing the land to wind storms and dust storms. Without trees to anchor the topsoil and slow the wind, soils are blown away.

An increasing population is widely thought to be the driving force behind desertification, but the relationship is not that simple. Boserup (1965) proposed the idea of 'induced innovation', as shown in Figure 6.4 — increasing populations and market opportunities create financial incentives for more intensive, yet sustainable land management. Machakos, a town in a semi-arid area in Kenya, experienced notable desertification in the 1930s. Yet today, despite the increase in population, this situation has been reversed as the farmers have learned to use the land in a sustainable manner — farming on terraces, using animal manure and compost on the land, and replanting trees (afforestation) to provide shelter belts. As markets develop and economic conditions improve, farmers can afford to buy fertiliser. Fruit, vegetables and milk are sold nationally, and coffee is grown for export. Equally, a decrease in population can lead to further desertification because of insufficient people to manage the land — this has been seen in villages in Chad with the loss of hillside terraces.

Figure 6.4 Induced innovation in arid lands

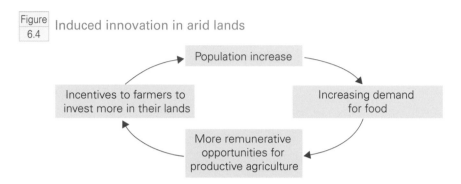

Poverty

Poor people, lacking access to the best land, often farm already fragile areas. Poverty means that they have little choice but to overexploit the land in order to solve economic problems in the short term. Invariably, the poor lack the resources to plan for the long term.

People living in poverty are the most likely to suffer the effects of desertification because they depend so much on the land for their livelihood. Yet removal of vegetation cover and soil nutrients leads to a decline in soil fertility, which in turn reinforces poverty. Nutrients are being steadily lost from African soils since many farmers cannot afford to use fertilisers. The protective cover of vegetation is also reduced as trees are cut down for firewood. As the land becomes less productive, farmers have to farm more land (often of poorer quality) until these lands in turn become degraded. Ultimately this leads to out-migration: with the land now useless, local farmers are forced to leave and swell the numbers of environmental refugees. It is estimated that 135 million people are at risk of being displaced by desertification. The problem is most severe in sub-Saharan Africa. It is estimated that 60 million people will move from sub-Saharan Africa towards northern Africa and Europe by 2020 as a result of desertification of their lands.

Where an economy is based on cash crops, the land is even more likely to be overexploited. Changing patterns of international trade can mean that local resources are overexploited for short-term export gains. One million hectares of land in Senegal, approximately half the cultivated area, is devoted to growing peanuts for export to Europe.

While there are strong links between poverty and land degradation, wealthy landholders can do even more damage to environmental resources. For example, tractor tillage by wealthier farmers removes soil cover on a much larger scale than is possible by hand or using animal tillage.

Policy errors and disasters caused by human activity

Relatively low priority may be given to environmental protection in LEDCs, with poor decisions being made about land management. In the past, only a limited amount of development assistance was given to dryland areas, since the return on investment was perceived as small. Mistakes in the choice of policies or technologies have led to land degradation in many countries, in both LEDCs and MEDCs.

The dustbowl in the American mid-west (Great Plains) in the 1930s resulted from a lack of knowledge of the environmental processes operating in the area. During a period of drought, ploughs that were designed for the more temperate climates of western Europe were used. Poor soil conservation methods led to severe erosion of the topsoil, creating clouds of dust. Improved management of agriculture and irrigation has since prevented another disaster, but because the mid-west is more suited to livestock grazing than to cultivation some soil erosion has continued to occur. In 1975, the US Council of Agricultural Science and Technology warned that severe drought in the Great Plains could trigger another dust bowl.

Large-scale dryland irrigation projects can cause salinisation, which in the worst cases can lead to the abandonment of the land. An example is the Aral Sea disaster. The Aral Sea, which was once the fourth largest lake in the world, shrank from a volume of approximately 1090 km³ in 1960 to around 162 km³ in 2000. This was a result of both high levels of water use from the lake itself and excessive water removal from the two feeder rivers (the Amu Darya and the Syr Darya) for irrigation. Current levels of irrigation are not sustainable. In 2007, the Aral Sea was just 10% of its original area. At the same time as depleting in volume, the water has become increasingly saline. In Kazakhstan, an ambitious US$68 million project financed by loans from the World Bank aims to regulate water usage and reverse the impact on the Aral Sea. A 13 km concrete dam, completed in 2005, has split the Aral Sea into two parts. The level of water in the North Aral Sea has already risen by 2 metres. The South Aral Sea, in Uzbekistan, is continuing to shrink because of a lack of management.

Unregulated access to land resources results in some individuals seeking to maximise their own gains by overexploiting the land. In north Africa, livestock farmers qualify for drought aid from their governments. The size of the monetary aid depends on the headage of livestock that the farmers own, which encourages larger herds and overgrazing.

Migration precipitated by wars and natural disasters such as floods and droughts can destroy productive land, which becomes overburdened by a heavy concentration of refugees. Conflicts in Darfur (Sudan), Eritrea, Ethiopia, Niger, Nigeria, Mali and Somalia have caused the mass movement of people from rural areas to refugee camps in nearby countries. This adds yet more pressure on the land as refugees move en masse into marginal ecosystems and carry on with their native farming practices, which may be unsuitable.

Activity 1

(a) Use the internet to acquire satellite images for the Aral Sea for every 4 years between 1960 and 2009. The websites **www.orexca.com/aral_sea.shtml** and Google Earth are good starting points. Trace the maps onto one piece of paper to show the change in size of the Aral Sea over this time frame.

(b) Describe the retreat of the Aral Sea since the 1960s.

(c) Why is the Aral Sea considered to be one of the world's worst ecological disasters?

(d) What could be done to slow down or reverse the retreat of the Aral Sea?

Salinisation

Salinisation is a form of land degradation in hot arid and semi-arid environments. When the water table is close to the surface, salts build up in the soil. This has a negative impact on crop production, as the salts are toxic to many plants.

Salinisation of cropland occurs predominantly through mismanagement. If excess irrigation water is used on the land, or if the soil is unable to drain rapidly enough, the water table will rise. The higher level of the water table means that the roots of crops or even the surface soil itself may be affected by salt water. If the water table rises to the surface, a salt crust forms, making the land useless for farming. Where salinisation is widespread and severe, farmers may be forced to leave the land.

Effects of desertification

Some deserts are separated sharply from less arid lands by mountains; elsewhere there is a gradual transition on the fringes of the desert. In the latter, the borders of the desert are not clearly defined and the transition zone supports a fragile and delicately balanced ecosystem. Geographically, deserts advance erratically. Semi-arid areas a long way from the borders of deserts can quickly degrade through poor land management. The causes of desertification are both natural and anthropogenic. It is difficult to separate them, and desertification in most areas results from the interaction of both.

According to the UNCCD, much of the arid and semi-arid agricultural area of the Earth has been harmed or is threatened by desertification, affecting around 250 million people in more than 110 countries. It is estimated that around US$42 billion worth of agricultural production is lost worldwide every year due to a combination of desertification and drought. In addition, approximately 41.5 million hectares of agricultural land loses all or part of its productivity every year. The UNCCD estimates that 70% of the world's drylands, covering an area nearly four times the size of Europe, are degraded and that 25% of the Earth's land area is to some extent affected by desertification.

Figure 6.5 shows the areas that are at greatest risk of desertification.

| Figure 6.5 | Areas at risk of desertification |

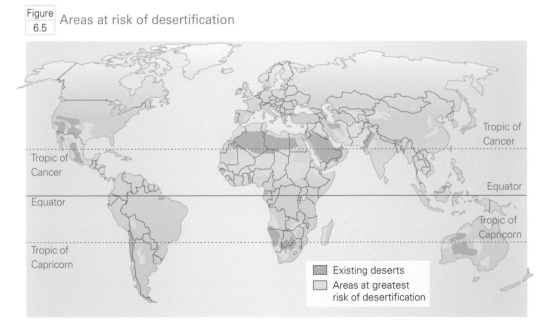

While hot arid and semi-arid areas in LEDCs suffer most from desertification, its effects can also be seen in MEDCs such as the USA, and especially in sub-humid grasslands like the African savannas. Desertification is becoming a problem in southern Europe as the Sahara spreads across the Mediterranean. A fifth of Spain's land area is already so degraded it is turning into desert.

Areas such as Sardinia and Sicily suffer from drought and tourism is making this worse — water tables fall as aquifers are overpumped to meet the soaring demand, leading to salt-water intrusion from the sea.

The impacts of desertification can be divided into economic, social and cultural, and environmental impacts, as shown in Table 6.2.

| Table 6.2 | The impacts of desertification |

Economic	Decline in agricultural fertility and productivity
	Reduced income from farming
	Less fuelwood available
	Greater dependence on food imports and food aid
	Increased poverty in rural areas
Social and cultural	Loss of traditional skills
	Increased frequency of food shortages and famine
	Forced migration because of food scarcity
	Social tension in the areas migrants move to
Environmental	Loss of soil nutrients as a result of water and wind erosion
	Removal of topsoil
	Decline in soil fertility
	Reduction in the amount of vegetation
	Reduced infiltration capacity, so less water in the soil
	Increased risk of flooding due to inadequate drainage or poor irrigation practices
	Loss of biodiversity where overgrazing leads to the loss of plant species
	Increased soil salinity, and the development of salt crusts
	Increased amounts of sand and dust
	Increased frequency of sandstorms
	Regional shifts in climate and alterations to the local energy balance

Desertification affects the local environment and the way of life of people living in these regions. There will be an impact in terms of loss of biodiversity, changing local climates and increasing demands on water resources. Those living in arid lands have to cope not only with existing deserts, but also with the threat of further desertification. The most immediate problem is the impact on agricultural areas. Desertification reduces the ability of the land to support life, including wild plants and animals, crops and domesticated animals, and people. As a process, desertification is self-reinforcing: once it has started, the land continues to deteriorate and the process is difficult to halt (Figure 6.6).

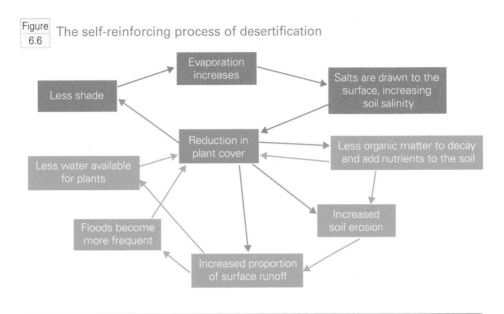

Figure
6.6 The self-reinforcing process of desertification

Activity 2

Study Figure 6.6. Explain why the process of desertification can be considered to be self-reinforcing.

Managing desertification

We have seen the seriousness of environmental damage caused by desertification and how it leads to a loss of resources in arid and semi-arid areas. Because it can take hundreds or thousands of years to regenerate soils, most of the damage is irreversible.

A range of management strategies have been used to protect the vegetation, soil and water resources in arid areas. Even so, economic and political pressures, combined with an increase in settled communities, population growth and a lack of emphasis on environmental protection, have resulted in increasing mismanagement of the land. Indeed, despite attempts in the 1970s and 1980s to tackle desertification, the United Nations Environment Programme (UNEP) concluded in 1991 that the problem of land degradation in arid, semi-arid and dry sub-humid areas had intensified, although there were local examples of success. Among the World Bank projects reviewed in 1985, almost one in three failed in west Africa, mainly because they did not work with traditional farming techniques.

In order to prevent further desertification, good management and a flexible approach is crucial. The practices that led to the desertification in the first place need to be changed. Simple technologies, such as soil conservation, reforestation and developing shallow groundwater reserves, can halt land degradation.

In 1996, the UNCCD (United Nations Convention to Combat Desertification) was established because of concern over the deterioration of land quality in more than 100 countries. It monitors desertification, and helps countries to draw up action plans to solve the problem by adopting a partnership approach between governments and local populations. Over 179 countries attended the 2002 UNCCD meeting and, by 2009, some 99 countries had drawn up action plans under the supervision of the UNCCD. While internationally respected agricultural scientists contribute to the plans, local and traditional expertise is considered important, and so farmers are consulted at a grass-roots level.

Activity 3

Use the website **www.unccd.int/actionprogrammes/menu.php** to find the action plan drawn up at national, regional and sub-regional levels for any country or area affected by desertification.

(a) Summarise the plan for your chosen country or area.

(b) Comment on how successful you think this plan will be.

Soil conservation is one of the keys to maintaining the biological productivity of the land. Table 6.3 shows some of the techniques that could be used to prevent or slow down desertification.

Table 6.3 Techniques that could be used to combat desertification

Problem	Possible solutions
Deforestation	Afforestation — replanting trees, especially in shelter belts
Soil erosion	Planting grasses can help stabilise the soil
	Re-seeding badly degraded areas
	Mulching to allow vegetation to re-establish
	Using cover crops of perennials or fast-growing annuals to protect the soil
	Wind breaks, such as trees planted at right angles to the prevailing wind
	Planting leguminous plants to extract nitrogen from the air and fix it in the soil
Overcultivation	Using good farming practices, e.g. crop rotation, using manure as fertiliser
	Crop rotation to help maintain productivity of the soil

Problem	Possible solutions
Overgrazing	Limiting grazing pressure by moving livestock to new areas at frequent intervals
	Fencing off areas to protect overgrazing by animals
	Reducing the number of animals on the land
	Introducing biodiversity action plans
Salinity	Improving irrigation techniques to reduce surface evaporation of water
	Installing salt traps (layers of sand and gravel) in the soil to prevent salts reaching the surface
Drought	Afforestation can help to reduce drought
	Terracing the land helps to slow surface runoff, and makes better use of rainwater
	Contour bunding — lines of stones along natural rises — helps to capture rainfall
Dune stabilisation	Conserving the plant community growing on the sides of dunes
	Planting xerophytic species to help slow/stop movement
	Erecting sand fences

The United Nations declared 2006 to be the International Year of Deserts and Desertification. Countries and international organisations were invited to increase public awareness of deserts and desertification during the year. Various activities celebrating the uniqueness of desert ecosystems and the cultural diversity of desert areas, and supporting the fight against land degradation, were promoted. The aim was to draw attention to desertification as a major threat to biodiversity, climate stability and human life.

Desertification in the Sahel

Location and overview

The Sahel in Africa comprises a transition zone between the dry Sahara Desert to the north and the wetter, more tropical south-central Africa to the south. This semi-arid belt stretches 3860 km across Africa from the Atlantic Ocean in the west to the Red Sea in the east (Figure 6.7). The width of the Sahel varies from several hundred to a thousand kilometres, and the total area is just over 3 million km². The semi-arid landscape is barren, with rock-strewn areas and stabilised ancient sand seas. Vegetation consists of xerophytic plants, including brush, grasses and stunted trees, that have adapted to cope with the low and unreliable rainfall.

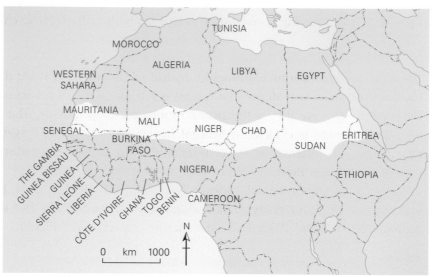

Figure 6.7 The location of the Sahel

The Sahel includes parts of nine African countries, and is home to more than 50 million people. It is one of the poorest and most severely degraded places on Earth, and has captured the attention of the media for over 40 years.

There is evidence that the Sahara Desert has undergone many periods of expansion and contraction. The evidence includes past distributions of sand dunes, changes in the levels of salt lakes, and documents from European explorers and settlers.

Since the mid twentieth century, the Sahara has expanded southwards into the Sahel at a rate of between 2 km and 5 km per year. The Director of the National Department of the Environment in Niger stated that an area of 2500 km², equivalent to the size of Luxembourg, is lost every year in Niger through desertification. There is no evidence that this advance will be permanent, however.

Climatic conditions

The average annual rainfall in the Sahel is low, varying from 200 mm to 600 mm. Rainfall is seasonal, and is concentrated in the summer months between June and September due to the movement of the ITCZ. However, annual rainfall figures can be misleading: even though there may be considerable amounts of rain in particular years, if the rain is confined to an intense burst over a short period of time it will bring little benefit to farmers. Moreover, high-intensity rainfall may actually destroy crops.

The Sahel is prone to drought. Large fluctuations between years of above-average and below-average rainfall are normal. Aridity, coupled with rainfall variability, creates a fragile environment that is highly vulnerable to change.

Periods of drought

The Sahel suffers from meteorological, hydrological and agricultural drought. For example, there is evidence that August is getting drier — a particular cause for concern because this is the prime month for crop growth. The impact of a drier climate is evident in Senegal in west Africa. There, irrigated rice production is threatened by lower rainfall and the subsequent reduction in flow of the Senegal River.

Figure 6.8 shows the rainfall index in the Sahel from 1950 to 2004. Rainfall was above average for many years during the 1950s and 1960s, but drought during the 1970s and 1980s had a devastating impact on the region.

| Figure 6.8 | Rainfall variability in the Sahel, 1950–2004 |

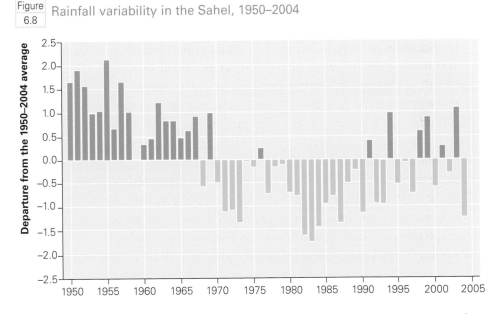

The droughts and famines between 1968 and 1974, and again during the period 1979–84, were widely reported. The media portrayed images of dry landscapes, dying cattle and emaciated, famished people. Over a million people died from starvation. At the height of the famine in the 1980s, an estimated 1000 people a month were dying, and 150 million people in 24 countries were on the brink of starvation. As many as 400 000 km² of land were laid waste, and it was thought that recovery would take a minimum of 50 years.

In June 2000, drought in the Sahel caught media attention once again (Figure 6.9). Six African countries faced devastation, with crops and livestock wiped out. Southeastern Ethiopia was the hardest hit. It received no substantial rain for 3 years. Ethiopia recorded 300 deaths in March 2000, mainly from diarrhoea, upper respiratory diseases and measles. However, the root cause of death was famine, with thousands weakened due to lack of food and water.

Figure 6.9 Drought in the Sahel

Activity 4

Use the internet to find online newspaper articles relating to the Sahel. In what ways have the media tried to portray the Sahel as inhospitable?

The following websites can be used as starting points:

www.timesonline.co.uk/tol/comment/columnists/guest_contributors/
    article548807.ece

www.guardian.co.uk/world/feedarticle/8461437

www.independent.co.uk/news/world/africa/millions-facing-famine-in-ethiopia-
    as-rains-fail-1779376.html

news.bbc.co.uk/1/hi/world/africa/710467.stm

news.newamericamedia.org/news/view_article.html?article_id=
    46b5712a3f4b1ef50a478654dc30dc83

Traditional life and agricultural activities

Traditional activities in the Sahel include nomadic herding and sedentary arable farming around permanent water sources such as rivers, lakes and oases. The indigenous peoples have adapted to the environment in a number of different ways:

- Pastoralist farmers are nomadic and move around to areas of seasonal vegetation growth; they avoid overgrazing of the land by moving on when pasture is in limited supply. Migration allows drylands that are not suitable for sedentary cultivation to be used sustainably.
- Farmers tend to leave vegetation around the more reliable or permanent water resources, rather than letting the animals overgraze. This means that they will be able to return to these areas later.
- Herd diversification, involving cattle, sheep and goats, is important. This allows pastoralists to use a wider range of the vegetation types as the different animals have different grazing/browsing habits. They also manage the size of herds to prevent overgrazing.
- Pastoralists barter their meat products with arable formers to obtain grain.
- More milk is consumed in wet periods, and more meat is consumed in dry periods.

Human causes of desertification in the Sahel

A natural green belt of bushes and trees extends across much of the Sahel and protects the region from the Sahara Desert to the north. However, this zone could easily be destroyed by cutting trees for firewood.

During French colonial times in west Africa (1881–1962), large-scale peanut cultivation was introduced, driven by the demand for vegetable oil in Europe. High market prices led to overcultivation and deforestation as the crop area was expanded.

Since the early 1960s, rapid population growth has occurred in the Sahel, with numbers rising from 19 million people in 1961 to 54 million in 2008. The CILSS (Permanent Interstate Committee for Drought Control in the Sahel) predicts that there will be 200 million people living in the region by 2050. Current population growth is 3% per year, compared to an annual growth rate of food production of just 2%. As the population has increased, traditional lifestyles have become increasingly sedentary and less nomadic. The impacts of rapid population growth and sedentarisation have been overcultivation, overgrazing, deforestation and the overexploitation of water resources. This unsustainable use of resources has resulted in desertification.

More specifically, a primary cause of desertification in the Sahel has been an expansion in the agricultural practice of slash and burn. Farmers degrade their environment by cutting and burning natural woodland in order to clear land for farming. The ash is scattered on the fields to provide nutrients to the soil. After harvest, the stalks of millet are cut and burned, exposing the fields to the strong winds. These winds blow away the topsoil, which, when deposited, suffocates plants and seedlings. Formerly the land was left fallow for periods of 7 years to recover before it was used again. With increasing population pressure, this is often no longer possible.

Solutions to desertification in the Sahel

After the droughts in the 1960s and 1970s, the international community sent much in the way of aid to the countries of the Sahel — including both emergency aid and long-term development aid. Most of this money was invested in water resource projects, in order to mitigate drought. Dams, irrigation canals and wells were built during the 1980s and 1990s. Most of the large dams (e.g. the Tiga Dam and the Kainji Dam) are located in the dry savanna areas of Nigeria, where the problems of water shortages and drought are most acute. Many of these projects were not successful, however. The reservoirs and irrigation systems were too small in scale to combat drought; they also brought new problems, such as providing breeding habitats for insects like tsetse flies, which transmit sleeping sickness. Some of the wells were dug in areas where the groundwater was not easily replenished, and actually contributed to desertification. Furthermore, because there were now more wells, the local people kept larger herds of livestock, which increased the problem of overgrazing. Trees were cut down to convert more land to agricultural use, altering the local albedo and precipitation patterns.

Mitigation of drought and desertification is possible, but rapid human population growth means that pressure on the land is likely to get more intense. There have been a number of solutions put forward to try to solve the problems of desertification and famine within the Sahel:

- On a local level, protection of local resources is encouraged by NGO projects such as conservation programmes for soil, water and agro-forestry.
- Improvements in the application of science and technology aim to make the land more productive — for example, the introduction of higher-yielding or drought-resistant crops and improvements to irrigation.
- Local farmers are encouraged to protect their own land through planting more perennial, drought-resistant species to protect the soil, as well as windbreaks to stop bare soil eroding.
- A famine early-warning system has been developed by USAID. This allows areas to predict the potential losses to crops and animals, and prevent food shortages caused by drought. Rainfall patterns are observed, and changes to vegetation cover are mapped.

- The Sahel and West Africa Club, which was set up by the international Organisation for Economic Cooperation and Development, offers help to areas within the Sahel.
- UK-based charity Oxfam offers short-term relief from food shortages. It has also developed programmes to help with long-term rehabilitation of the Sahel.
- In 1973 the countries in the Sahel region joined together to form an alliance to combat the severe drought — this was the CILSS (Comité permanent Inter-Etat de Lutte contre la Sécheresse dans le Sahel, or Permanent Inter-State Committee for Drought Control in the Sahel). The aim was to 'invest into the research for food security and in the fight against the effects of drought and desertification in order to achieve a new ecological equilibrium'. It works to unite the different regional efforts.

The Green Wall Sahara Initiative

The Green Wall Sahara Initiative, approved in 2009, includes a US$3 million, 2-year phase to plant a belt of trees from Mauritania to Djibouti. The belt will be 7000 km long and 15 km wide. The aims of the scheme are to:
- arrest soil degradation and reverse desert encroachment
- increase land productivity in more than 25 countries
- protect cultivated land from wind erosion
- conserve biodiversity
- promote sustainable land management
- provide a sustainable timber source to the local population
- improve livelihoods by creating millions of jobs
- promote ecotourism
- create public awareness of the fight against desertification

Diguettes in Burkina Faso

One of the most successful local schemes to combat desertification is the building of diguettes (contour stone lines) in Burkina Faso.

Local farmers used a traditional technique of piling heavy stones in the path of water as it flowed downhill. While this worked to slow surface runoff following rainfall, there were some problems. Because the walls didn't follow contour lines, water piled up at areas of low relief and broke through the stone walls, causing erosion.

After the drought in 1973 and 1974, Oxfam worked with farmers to increase food production by improving the traditional stone walls and making them more effective. Training sessions helped to teach farmers how to build the walls along the natural contours of the hills. By aligning the diguettes in this way, water is dammed uphill, allowing it more time to sink into the earth. This slows down soil erosion, and increases the amount of water in the soil for use by crops. It also encourages the deposition of nutrient-rich sediment on the farmland. Diguettes are now used by hundreds of villages, and can be seen almost everywhere in Burkina Faso (Figure 6.10). Using this system, the land is more productive; millet production has increased by 50% on average.

Figure 6.10 Building diguettes in Burkina Faso

Mark Edwards/Still Pictures

The project was successful for a number of reasons:
- It used free and easily available local materials.
- It cost nothing to construct the walls.
- While labour intensive, the work to build the walls can be done in the dry season.
- The walls are simple to make and maintain.
- They are based on an improvement of an existing technology.

Activity 5

Using the internet to help you, research an NGO that is working in the Sahel, such as the Eden Foundation, Action Against Hunger, CARE International or Oxfam. Make notes on how they are trying to help local farmers to control desertification.

7 Managing economic development in arid environments

Historically, the inhospitable environment and harsh terrain of arid areas meant low population densities. Today, however, many arid areas are being developed to exploit the richness of their resources. They have become important locations for economic activities such as agriculture, tourism, recreation, mineral extraction and film making. This economic potential has led to the growth of urban areas in many arid regions. Consequently, while arid and semi-arid environments now provide significant new opportunities, their inherent fragility calls for careful management.

Water resources are key to sustainable economic development in these regions. In the past, the over-extraction of water resources has compromised development. In order to achieve sustainable development, a balance between socio-economic and environmental needs is essential.

Agriculture

Over thousands of years, a nomadic lifestyle has evolved in many arid and semi-arid areas as a way of coping with the scarcity of water. Pastoral nomads include the Tuareg in southern Algeria, the Bedouins of northeast Africa and the Arabian Peninsula, and the Somali nomads. Sheep, goats, donkeys, horses and camels have been exploited since the Neolithic revolution (*c.* 8000–6000 BC) for milk, meat, hides, wool and for transport. However, indigenous groups are increasingly abandoning the nomadic way of life and instead are settling in towns and villages. Elsewhere a few indigenous peoples, such as some Australian Aborigines, still lead a hunter-gatherer existence.

Agriculture in arid and semi-arid areas might seem unlikely, but abundant sunshine and a continuous growing season can create ideal conditions for cultivation. However, there is one proviso: irrigation water must be available. A variety of water sources occur within arid and semi-arid areas, and these are outlined in Chapter 2. Advances in irrigation techniques have meant that many crops are now grown for export. It is important to continue to develop more efficient ways to manage irrigation to conserve water. The consequences of overexploitation of water resources and poor land management leading to desertification and land degradation are discussed in Chapter 6.

Farming without irrigation

The Hopi, Navajo and other North American peoples in the southwest USA practised desert farming without irrigation. They grew corn in washes (seasonal flood plains), valleys between mesas, and on terraces along mesa walls that naturally collect water.

The dry-farming systems in the savannas of north and west Africa still survive today. Dry farming is much more widespread in this area than irrigation agriculture. For example, maize is grown without irrigation on the Atlantic coast of Morocco; extensive olive groves are located on the sandy soils at Sfax in Tunisia, despite an annual rainfall of only 254 mm. Runoff is checked by dams made of stones or brushwood in order to retain moisture for a longer period and allow the water more time to sink into the soil. A system of hillside terracing arranged to reduce the speed of runoff down the slope works in a similar way. But dry farming is a risky activity, dependent on uncertain natural rainfall.

Traditional irrigation techniques

Irrigated farming in arid and semi-arid regions has been carried out since ancient times. Advances in irrigation techniques were central to the Neolithic revolution. It has been argued that these innovations were a response to the climatic deterioration that occurred as the ice sheets in the northern hemisphere retreated.

Oases and rivers provide opportunities for irrigated agriculture. For example, along the banks of the Draa River in Morocco, which flows from the Atlas Mountains through arid lowlands towards the Sahara Desert, crops such as dates, citrus fruits, barley and vegetables are cultivated. Water is transferred to the fields using simple dykes built from earth or stone (Figure 7.1). Dykes can also be used to divert the floodwaters from wadis onto croplands.

In ancient Egypt, the floodwaters of the Nile were diverted into canals and used for irrigation, and water was also raised from wells. Water from aquifers was brought up to the surface using lifting devices (such as the Archimedes screw) powered by either humans or animals. Similar methods of raising water from wells are still used in parts of Africa today.

Figure
7.1
Dykes to transfer irrigation water, Morocco

Michael Raw

Figure
7.2
Schoolchildren collecting water, Zimbabwe

Neil Cooper/Still Pictures

In ancient Persia (now Iran), **qanats** supplied a regular flow of water for irrigation. They are among the oldest known irrigation methods and are still used today. Similar structures are found in other countries, including Afghanistan, Pakistan, China, the countries of the Arabian Peninsula, Egypt, Libya, Algeria, Tunisia and Morocco. A qanat is a gently sloping underground channel, typically 50–100 cm wide and 90–150 cm high, for bringing water from an underground aquifer (e.g. in the foothills of a mountain) to an artificial oasis some kilometres distant. This has the advantage that water passing through the underground channel is not subject to evaporative loss. Wells at various points sunk into the underground passage allow local water extraction en route. The isolated oases found along the Draa River Valley in Morocco close to the Sahara Desert rely on qanat water from the foothills of the Atlas Mountains for irrigation. Locally they are known as 'khettara' (Figure 7.3).

<table>
<tr><td>Figure
7.3</td><td>Diagram of a typical khettara in Morocco (not to scale)</td></tr>
</table>

(a) Cross-section

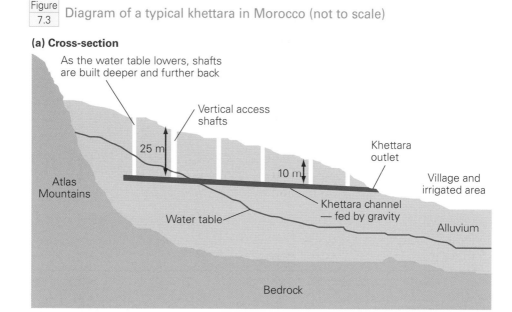

(b) Aerial view

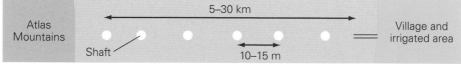

Modern irrigation techniques

While modern irrigation techniques may be more efficient and waste less water, the high capital outlay required for their implementation places limits on their availability to poor farmers and smallholders. Modern irrigation techniques fall into two main areas:

- constructing dams on a large scale to collect and store water
- developing high-tech methods to irrigate crops with minimum wastage of water

Dams

In modern times, dams on major rivers have facilitated large-scale irrigation of arid regions. For example, the Colorado River flows 2330 km from its source in Colorado to the Gulf of California, passing through the arid region of the southwest USA. Water supply for the states of Colorado, Arizona, Utah, Nevada and New Mexico comes mainly from the Colorado and its tributaries. Dam construction on the Colorado River to provide water for agricultural and urban use began in the early twentieth century. The two largest dams are the Glen Canyon Dam and the Hoover Dam (see Figures 7.4 and 7.5). Water is stored behind the dams in the Lake Powell and Lake Mead reservoirs respectively. Irrigation water provided by the reservoirs supports intensive agriculture in the dry zone of Arizona, Nevada and California. At the Imperial Dam, massive amounts of water are diverted through the All-American Canal to the Imperial Valley in southern California — a desert region that lies below sea level and has an annual rainfall of only 7 cm. The All-American is the largest irrigation canal in the world and can carry 850 m³ of water per second. Thanks to water from the Colorado River, the Imperial Valley is now an important agricultural region, producing cotton, grain, and winter fruits and vegetables.

| Figure 7.4 | Glen Canyon Dam, Arizona |

Lucy Cole

Water extraction from the Colorado River is regulated by the Colorado River Compact, which allocates a percentage of the water to each state (see Table 7.1). There is current debate over the water allocations, and concern that excessive water abstractions (mainly for irrigation agriculture) have lowered

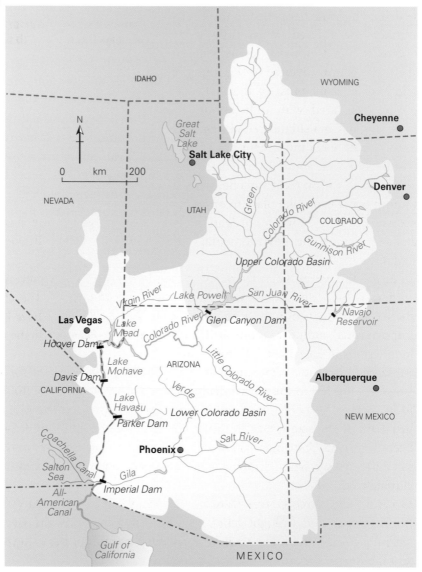

Figure 7.5 The Colorado drainage basin: dams and reservoirs

the water table over wide areas. Other concerns are that the river no longer consistently reaches the sea in the Gulf of California, and its salinity increases significantly downstream. Dams are also criticised for wasting water, since large amounts are lost through evaporation and seepage.

Before the dams were built, the Colorado had one of the largest delta estuaries in the world. However, for a period of 6 years in the 1930s (as Lake

Mead filled following the completion of the Hoover Dam) no fresh water reached the delta. The same happened in the 1960s as Lake Powell filled. What little water remains in the Colorado downstream of Lake Mead is now diverted to the Imperial Valley. As a result, most of the delta has become desiccated with only about 5% of the original marshland remaining. The energy cost of water movement can also be considerable, with about 20% of electricity consumption within California being used to pump water from one place to another.

| Table 7.1 | Water allocation from the Colorado River by state |

State within drainage basin	Water allocation (billion m³)
Colorado	4.79
Wyoming	1.30
New Mexico	1.04
Utah	2.13
California	5.43
Arizona	3.70
Nevada	0.37

Construction of dams for irrigation and electricity generation has to be balanced against the environmental costs, which are often heavier than originally supposed.

- In Egypt, the Aswan High Dam, completed in the 1960s, has prevented silt carried in the Nile's floodwaters from reaching lower Egypt. The environmental consequences have been severe: soils in the Nile Delta have become impoverished and now require artificial fertiliser, and fisheries in the eastern Mediterranean have been adversely affected.

- In central Asia, the redirection of rivers feeding the Aral Sea for cotton irrigation has led to the virtual disappearance of what was once the world's fourth largest body of inland saline water (see Chapter 6). The Aral is now just 10% of the size it was in the 1960s, and its once-thriving fishery has been destroyed. There has also been heavy pollution caused by wind-blown salt and fertiliser, while drinking water has been contaminated and the local climate changed.

- In China, the Three Gorges Dam on the Yangtze River, though providing flood control and drought relief, prevents silt carried by the river from reaching coastal areas. The result is accelerated erosion of the Yangtze Delta and the loss of fisheries.

Where rivers cross international boundaries and water resources have to be shared, conflict can arise. For example, water from tributary rivers in Jordan and Syria flows into the Sea of Galilee in Israel and then into the River Jordan. Plans to divert the flow of these tributaries were a major cause of the 1967 war between Israel and its neighbours. Similarly, the headwaters of the Tigris and Euphrates rivers are in Turkey, while their lower basin is in Syria and Iraq.

Construction of major dams in Turkey in the 1980s and 1990s, such as the Atatürk Dam, has reduced the flow in the Tigris and Euphrates and threatens agriculture, especially in Iraq.

Activity 1

(a) Using the data in Table 7.1, draw a pie chart to represent the proportion of water allocated by state.

(b) On an outline map of the Colorado drainage basin, use proportional symbols to show the amount of water allocated in each state.

(c) Discuss the potential for conflict regarding water allocation on the Colorado. The following websites are useful starting points:

www.america.gov/st/webchat-english/2009/May/20090519132309HMnie
  tsuA0.6564905.html

www.ag.arizona.edu/AZWATER/arroyo/101comm.html

www.infocusmagazine.org/7.1/env_colorado_river.html

www.santafenewmexican.com/Local%20News/More-conflict-expected-over-
  Colorado-River

Economy in the use of water

At farm level, more efficient methods of irrigation have provided economy in the use of water and a more natural irrigation pattern.

Advances in electric motors to pump groundwater mean that in many arid areas, despite more efficient methods of irrigation, water is extracted faster than it is replaced by rainfall.

Water may be taken from aquifers containing fossil water that has remained in the ground for millennia, as with the Ogalalla Aquifer in the USA and the Nubian Sandstone Aquifer System in the Sahara. The Nubian and Saudi Arabian aquifers consist almost entirely of fossil water, between 2 million and 10000 years old, and are only recharged very slowly under present-day climatic conditions. Current water yields are unsustainable and eventually the reservoirs will be depleted. The scale of water extraction is evident in Saudi Arabia, where centre pivot irrigation from underground aquifers permits the growth of wheat and alfalfa. Annual groundwater withdrawals in Saudi Arabia are approximately 15 times greater than the recharge.

Activity 2

Food produced in hot arid countries in the developing world using irrigation is exported to many MEDCs. Discuss the ethics of this.

| Table 7.2 | Modern methods of irrigation |

Method	Description
Rapid flood	Water levels on the field are controlled by dykes. The water is pumped to the level of the land to flood cultivated fields using a controlled amount of water. This is used for crops that need to be occasionally submersed in water, rather than those that require gradual water inputs.
Sprinkler	Water is piped to central locations within the crop area. High-pressure sprinklers are mounted overhead or on moving platforms. This is a more economical method of using water than flood irrigation.
Centre pivot	Sprinkler pipes are mounted on a wheeled tower, which typically rotates through 360° once every 3 days. This provides a natural and evenly distributed pattern of moisture mimicking natural rainfall and low-energy precision irrigation (LEPA).
Drip	A very localised form of irrigation: water is delivered to the root zone of the plants drop by drop using buried plastic pipes. It is a slow and continuous method of watering plants. Developed in Israel in the 1950s, it is considered to be the most valuable contribution to irrigation technology; it provides an efficient, precisely controlled flow of water to the roots of the growing plants, and conserves water by minimising any loss through evaporation, runoff and deep drainage.
Hydroponics	This is the use of mineral nutrient solutions to grow plants without soil. Crops of a high value are grown in containers in greenhouses. The largest commercial hydroponics area in the world, owned by Eurofresh Farms in Arizona, USA, extends over 1.29 km².

Activity 3

Construct a concept map to show (a) how agriculture can be managed sustainably and (b) the limits to the sustainable development of agriculture in LEDCs.

Tourism and recreation

High temperatures throughout the year, plenty of sunshine and a low likelihood of rainfall mean that arid areas provide an ideal holiday environment. Tourism in arid areas covers a range of different activities:

- Many arid and semi-arid coastlines have been developed as resort areas, capitalising on beach tourism — for example, resorts along the coastlines of some states of the Persian Gulf.
- Much of the desert tourism in the southwest USA is focused on viewing dramatic desert landscapes and scenery. The national parks of the southwest,

such as Grand Canyon, Bryce Canyon, Arches (Figure 7.6), Joshua Tree and Zion, attract millions of visitors a year.

- Spectacular wildlife in arid and semi-arid regions attracts visitors — for example, black rhinos, elephants and lions in Namibia.
- Cultural tourism to Egypt is long established and is based on the ancient Egyptian civilisation and archaeological sites. The pyramids attract millions of visitors a year.
- Extreme sporting events, such as cycle races across the Sahara Desert and white-water rafting along the Colorado River, cater to a specific tourist demand.
- Warm seas have led to the development of a range of water sports activities, most notably snorkelling and scuba diving in coastal areas supporting coral reefs, for example at Sharm el-Sheikh, Egypt.

Figure 7.6   Double Arch, Arches National Park, southwest USA

Michael Raw

Tourism provides tangible economic benefits for many countries in the arid and semi-arid zone, such as foreign currency earnings and employment. The growth of tourism encourages investment in infrastructure, so roads, railways, airports, hotels, gas, electricity and water supplies are all built or improved. Yet tourism also brings with it a set of challenges. These fragile areas have a low carrying capacity, and are highly vulnerable to human activity. An increase in the number of visitors inevitably leads to some environmental damage, degrading natural landscapes, habitats and ecosystems. Off-road vehicles damage fragile soils, cause stress to wildlife and contribute to air pollution.

Problems are most severe at popular destinations where there is overcrowding and unsustainable use of resources.

Las Vegas and Palm Springs exemplify inland resorts in the desert. In urban settlements in the southwest USA, most water use is in homes, gardens, industry and agriculture. However, the recreational pursuits associated with tourism increase water use. For example, in the Coachella Valley, which includes Palm Springs, the 120 golf courses consume 17% of all water used. The 57 golf courses near Las Vegas account for 7.6% of local water consumption.

Some forms of desert-area tourism, such as the high-adrenaline sports of desert safari, including dune bashing, wadi bashing, sand skiing and sand boarding, are more directly destructive of the environment. Dune bashing and wadi bashing involve driving off-road vehicles such as four-wheel drives or quad bikes at high speed around sand-dune environments. Such activities are becoming increasingly attractive in parts of the Middle East, such as Dubai, and have an adverse impact on the wildlife and environment.

Tourism does not just threaten the natural environment: it also has social and cultural disadvantages. Indigenous cultures may become commercialised for the tourist market, leading to a loss of local traditions. Tourists may cause offence to locals by dressing inappropriately or through behaviour that shows a lack of respect for places of spiritual importance to local people. For example, some Native Americans are not happy with the development of the Grand Canyon Skywalk, as they consider the Grand Canyon to be sacred ground. Local villagers may be forced to move as a result of tourism developments — for example, villagers living near Luxor in Egypt were forced to move to prevent household waste water from contaminating local tombs. The establishment of national parks, such as Etosha in Namibia, restricts the access of indigenous peoples to resources and nomadic grazing areas.

Sustainable tourism

Careful management is needed to ensure the sustainability of tourism developments within fragile arid and semi-arid environments. The aim of sustainable tourism is to help to generate jobs and income for local people, while at the same time protecting local customs and environments. Tourism can be managed sustainably by:
- creating national parks to protect landforms and ecosystems
- establishing conservation areas
- closing conservation areas to private vehicles and instead providing buses as transport

- controlling visitor numbers
- using information boards and visitor centres to educate tourists about the local environment, and how they can help to protect it
- establishing 'honeypot' sites to limit the damage (by concentrating tourists in a particular area, fragile ecosystems further away are protected)
- establishing tourist codes of conduct
- building tourist facilities in keeping with the local area
- using money generated from tourism to restore historical sites

Desert tourism in areas such as Morocco takes a more 'ecotourism' approach. Visitors trek into the desert on camels (Figure 7.7) and camp there, living as the Bedouins and Berbers have for centuries (Figure 7.8). This causes minimal damage to the natural environment.

Figure 7.7 Tourists in Morocco travelling into the Sahara Desert

Lucy Cole

Figure 7.8 Tourist accommodation: a traditional Berber camp in the Sahara Desert

Lucy Cole

Activity 4

Study Figure 7.6 on page 116, a photograph taken in Arches National Park in the southwest USA.

(a) What are the opportunities and threats to the environment posed by encouraging tourism to desert sites such as this?

(b) To what extent is it possible to manage such areas sustainably? Include reference to the management of Arches National Park (**nps.gov/ARCH/index.htm**).

Sharm el-Sheikh, Egypt

Many arid and semi-arid coastal regions have warm seas that encourage tourism. In north Africa, tourist resorts extend from Morocco to the Red Sea coast in Egypt. Between 1970 and 2009, Sharm el-Sheikh, at the tip of the Sinai Peninsula (Figure 7.9), grew from a small fishing village with about 100 inhabitants to a city with a population of 35 000. Its major industry is tourism, and it has become popular for various water sports, most notably scuba diving and snorkelling. Tourist arrivals increased from 8000 in 1982 to 1.2 million in 2004. However, an over-reliance on tourism can be risky, particularly in countries that are politically unstable. The terrorist attack on Sharm el-Sheikh on 23 July 2005 caused a temporary decline in tourism to Egypt and to Sharm el-Sheikh in particular.

| Figure 7.9 | The location of Sharm el-Sheikh |

Because of their clear water and their biodiversity, the coral reef ecosystems at Sharm el-Sheikh are among the best in the world. Since 1983 they have been protected as part of the Ras Mohammed National Park, which covers the entire shoreline fronting the Sharm el-Sheikh tourism area. The park is managed to protect the coral reefs from degradation and from the threats posed by tourism and development activities. Divers are encouraged to recognise their individual responsibility, by not touching the coral or collecting souvenirs. The Gulf of Aqaba Protectorates Development Program from 1997 to 2001 aimed to establish ways to conserve the natural resources of the area and protect the reefs for future generations of divers, as well as to achieve sustainable economic development of tourism activities.

Mining and mineral wealth

The presence of valuable fossil fuels such as oil and natural gas in the Middle East has been instrumental in transforming these nations' economies. Oil and gas are found in Algeria, Libya, Saudi Arabia, Kuwait, Bahrain, Qatar, the United Arab Emirates (UAE), Iraq and Iran. Of the word's readily accessible oil reserves, 62.5% are located in Saudi Arabia, the UAE, Iraq, Qatar and Kuwait.

Other important mineral resources in arid areas include:
- copper in the Atacama Desert, Chile
- diamonds and uranium in the Kalahari Desert, Namibia
- salt in the Sahara Desert, Niger
- gold in the Tanami Desert, Australia
- opals in the South Australian desert

Mineral reserves in arid and semi-arid environments are often located in remote areas, and present particular challenges for development and exploitation. Mining requires roads and railways in order to transport the minerals for processing or export. Miners require housing, water supplies and services. Mining towns have grown up in areas that are otherwise inhospitable. For example, much of Australia is underlain by ancient crystalline bedrock, rich in minerals. This mineral wealth has been extensively exploited in recent decades, leading to the opening up of the Outback (Figure 7.10). There are still nearly 100 working gold mines on the Darling Plateau in Western Australia, and opal mining is a major activity in South Australia and New South Wales. In the Pilbara region of Western Australia, mining of iron ore is a major earner. Iron ore resources are estimated at 32 billion tonnes, and 90% of the annual production is exported to the integrated steel markets in Asia. Strong Chinese demand has encouraged the development of new mines.

Figure 7.10 Mining in the Australian deserts

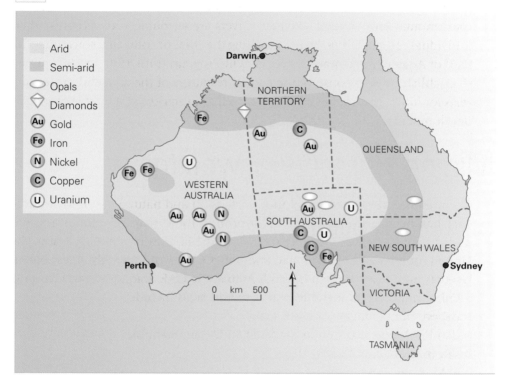

Fluctuations in global demand and prices for minerals, however, mean that mining can be a risky business. The unpredictable nature of mining is evidenced in the many 'ghost towns' found in desert areas — towns that used to house miners and their families but have since become abandoned and deserted. Some of these ghost towns are now tourist attractions. Coolgardie in Western Australia, a former gold-mining town and the original site of the gold rush to Western Australia in the 1890s, today contains museums and exhibitions depicting the equipment and methods of mining used in its heyday.

Mining also has negative impacts on the environment. Modern mining requires large quantities of water, which is in short supply in desert areas. The construction of roads and processing plants destroys desert vegetation and damages soils. Opencast mining and spoil heaps create eyesores. The spoil heaps are also a source of dust that can be blown by the wind and pose health problems — for example, spoil heaps from the borax mines near Death Valley have a low toxicity. Liquid waste from ore treatment plants and seepage from treatment ponds can contaminate the local water and soils. In many areas today, mining companies are obliged to carry out environmental assessments to reduce their impact on the area in which they operate, and to restore the environment once mining has ceased.

Activity 5

Study Figure 7.10, which shows the location of mineral reserves within the arid areas of Australia. Investigate the potential opportunities and challenges associated with mining in such areas. The following websites can be used as starting points:

www.sea-us.org.au/ozimpacts.html

www.aph.gov.au/Senate/Committee/uranium_ctte/report/d03.htm

www.atse.org.au/index.php?sectionid=215

Oil wealth in Dubai

Dubai, situated in the Persian Gulf, is one of the seven states that make up the United Arab Emirates (Figure 7.11). Before the discovery and export of oil in 1966, the nation had a simple subsistence economy. The interior was inhabited mainly by nomadic pastoralists, and Dubai was a small port town. The main source of income was from pearl fishing; the town also acted as a trading centre between India and Iran.

The money that the oil has generated has led to rapid economic development and industrialisation, and Dubai has been transformed into a modern city. The population has increased tenfold, mainly through foreigners migrating temporarily or permanently to the city. However, there are two questions that can be asked about the sustainability of development in Dubai.

First, how will Dubai support itself when the oil runs out? It is estimated that its oil reserves will be exhausted by 2025, yet with over 90% of the territory of Dubai classified as desert there are few other opportunities for economic development. In fact, unlike the situation in many of the other countries of the Persian Gulf, in 2004 only 17% of the money earned by Dubai came from oil — compared to 93% in Kuwait and 60% in Bahrain. Over the last 20 years, Dubai has been actively trying to diversify its economy, and its three biggest successes have been in tourism, IT, and banking and commerce. Dubai has a mixed state and private economy, which allows foreign banks and investors to operate there, attracted by the excellent infrastructure, liberal government,

Figure
7.11

The location of Dubai, UAE

free-trade zone and lack of taxes. Dubai International Airport makes Dubai highly accessible to tourists, and there were 7.5 million visitors in 2008. Along with beaches, coral reefs, historic sites and duty-free shopping, attractions include developments on a grand scale such as the Emirates Towers (the 12th and 24th tallest buildings in the world), the Burj al-Arab (the world's tallest building and most expensive hotel, located on its own island in the Persian Gulf) and the Jumeirah Palm (the world's largest artificial island, shaped like a palm tree).

Second, what is the impact on the water supply? Dubai lies within the Arabian Desert, and therefore gets little rain. Water consumption has increased with development, and the UAE today has the highest water consumption per person in the world. The growth of tourism has increased the demand for water for hotels, swimming pools and golf courses. As groundwater reserves have been depleted, Dubai has relied increasingly on desalinisation of seawater. Approximately 70% of the UAE's water comes from desalinisation plants, and Dubai has an installed desalinisation capacity of 188 million gallons (855 million litres) per day. Dubai has six desalinisation plants, with a further five either under construction or in planning phase, and aims to invest US$20 billion over the next 8 years to boost its capacity by over 2000 million litres of water a day.

The film industry

Deserts contain a variety of dramatic landscapes, which provide an ideal location for Westerns and for science-fiction, historical and Biblical films. Clear skies, plenty of sunshine, and extremes of light and shadow also prove advantageous when shooting films.

Film making can bring many advantages to the local economy. Fees are charged to the film companies for using the locations. Local people are employed to work as extras, while local services such as hotels and catering companies will be used by the crew. Although the benefits to the economy during filming are short term, there can be long-term benefits. The money generated can be used to improve the infrastructure, making the area more accessible, and the setting of the film may also encourage tourism to the region.

The Atlas Film Corporation, Ouarzazate

Ouarzazate, in Morocco, has developed as an important centre for the film industry, and the Atlas Film Corporation has studios there. There is guaranteed sunny weather and labour costs are low. Added to this, distinctive scenery and the availability of 'authentic' Moroccan locals mean that this is a particularly good location for movies involving historical storylines and desert settings.

Ouarzazate has been the location for several blockbuster hits, including:

- *Lawrence of Arabia* (1962)
- *Star Wars* (1977)
- *The Mummy* (1999)
- *Gladiator* (2000)
- *Asterix and Obelix: Mission Cleopatra* (2002)
- *Babel* (2006)

The film studio itself brings tourism to the area, offering guided tours of the various movie sets spread over 150 hectares.

There are plans to spend £3.2 million to turn the city into an international film centre by 2016. Improving local resources would enable Ouarzazate to increase production from 11 to 38 films per year. According to the Moroccan Cinema Centre, this could generate an annual income of £153 million for Morocco and create 8000 new jobs.

Figure 7.12 shows the nearby Aït Benhaddou, which is a ksar (fortified city) and UNESCO World Heritage Site. This traditional Moroccan city was established in the eleventh century, and is located 30 km outside Ouarzazate.

Lucy Cole

Figure 7.12   Aït Benhaddou

An estimated 130 000 tourists visit Aït Benhaddou every year. The first major film to feature Aït Benhaddou was *Lawrence of Arabia* (1962). More recent films shot here include *Gladiator* (2000) and *Alexander* (2004). Its advantages as a movie location have helped to prevent this site from falling into total disrepair. It is constructed using mud bricks, which makes it vulnerable to the elements and necessitates regular restoration. In 1977 its lower façades were restored for the filming of *Jesus of Nazareth*. There is currently a restoration programme in place aiming to prevent further erosion of the ksours (mud houses) and eventually to repopulate the town.

Scientific research

Specific scientific research is carried out on deserts in order to increase knowledge about this environment. General scientific research is also carried out within deserts, taking advantage of the remoteness, sparse population, and natural conditions that lead to a high visibility and excellent preservation of rocks and materials. Desert regions have a long history of being involved in scientific research, and much of this has had a significant impact on global science, space research and the defence industry.

- In the 1950s and 1960s, the UK based much of its rocket and nuclear testing in the South Australian desert.

- Early atomic bomb tests were carried out in arid areas such as New Mexico.
- The deserts in the southwest USA provide sites for military and research bases.
- Research into nuclear energy and nuclear weapons takes place in Los Alamos, New Mexico.
- Many astronomical observatories are placed in deserts, where the cloudless skies and lack of artificial lights are advantageous — for example, Mount Wilson Observatory in the Californian desert, where the renowned astronomer Edwin Hubble worked.
- NASA carried out experiments in the deserts of the southwest USA to test communication rovers and robot interactions, as the barren conditions were considered similar to those on the Moon and Mars.
- Most meteorites that have been collected for study come from deserts, where they are well preserved because of the slow rates of erosion.

Activity 6

Use the website **www.galeschools.com/environment/biomes/desert/human.htm** to complete the following tasks:

(a) Describe the ways in which nomadic peoples traditionally adapted to desert conditions.

(b) Examine the impact of humans on the desert.

Activity 7

(a) Construct a concept map to show the main obstacles to economic development in arid lands.

(b) Explain one way in which each obstacle to development could be overcome.

(c) Investigate a specific arid region or country, and assess how successful people have been in overcoming the obstacles to development. To what extent is the development of this area sustainable?